Lecture Notes in Electrical Engineering

Volume 865

The book series *Lecture Notes in Electrical Engineering* (LNEE) publishes the latest developments in Electrical Engineering - quickly, informally and in high quality. While original research reported in proceedings and monographs has traditionally formed the core of LNEE, we also encourage authors to submit books devoted to supporting student education and professional training in the various fields and applications areas of electrical engineering. The series cover classical and emerging topics concerning:

- Communication Engineering, Information Theory and Networks
- Electronics Engineering and Microelectronics
- Signal, Image and Speech Processing
- Wireless and Mobile Communication
- Circuits and Systems
- Energy Systems, Power Electronics and Electrical Machines
- Electro-optical Engineering
- Instrumentation Engineering
- Avionics Engineering
- Control Systems
- Internet-of-Things and Cybersecurity
- Biomedical Devices, MEMS and NEMS

For general information about this book series, comments or suggestions, please contact leontina.dicecco@springer.com.

To submit a proposal or request further information, please contact the Publishing Editor in your country:

China

Jasmine Dou, Editor (jasmine.dou@springer.com)

India, Japan, Rest of Asia

Swati Meherishi, Editorial Director (Swati.Meherishi@springer.com)

Southeast Asia, Australia, New Zealand

Ramesh Nath Premnath, Editor (ramesh.premnath@springernature.com)

USA, Canada:

Michael Luby, Senior Editor (michael.luby@springer.com)

All other Countries:

Leontina Di Cecco, Senior Editor (leontina.dicecco@springer.com)

**** This series is indexed by EI Compendex and Scopus databases.****

More information about this series at https://link.springer.com/bookseries/7818

Zahriladha Zakaria · Seyed Sattar Emamian
Editors

Recent Advances in Electrical and Electronic Engineering and Computer Science

Selected articles from ICCEE 2021, Malaysia

Springer

Editors
Zahriladha Zakaria
Faculty of Electronic Engineering
and Computer Engineering
Centre for Telecommunication
Research and Innovation
Universiti Teknikal Malaysia Melaka
Melaka, Malaysia

Seyed Sattar Emamian
Sigma Research and Consulting
Delft, Zuid-Holland, The Netherlands

ISSN 1876-1100 ISSN 1876-1119 (electronic)
Lecture Notes in Electrical Engineering
ISBN 978-981-16-9780-7 ISBN 978-981-16-9781-4 (eBook)
https://doi.org/10.1007/978-981-16-9781-4

This Springer imprint is published by the registered company Springer Nature Singapore Pte Ltd.
The registered company address is: 152 Beach Road, #21-01/04 Gateway East, Singapore 189721, Singapore

Preface

This book presents a compilation of research works covering the fields of Computer Science, Electrical and Electronic Engineering. All the manuscripts in this volume were presented during the 3rd International Conference on Computer Science, Electrical and Electronic Engineering 2021 (ICCEE 2021) which was conducted through a virtual presentation on 18 August 2021. This conference holds a vital role as a catalyst in seeking wisdom, sharing thoughts and opinions to promote better understanding of engineering and information technology and making the leap for the green technologies.

The editor(s) of the proceeding would like to express their utmost gratitude and thanks to all reviewers in the technical team for making this volume a success.

Melaka, Malaysia — Prof. Dr. Zahriladha Zakaria
Delft, The Netherlands — Dr. Seyed Sattar Emamian

Acknowledgements

The editors would like to thank all the members of the local organizing committee who helped organize the 3rd International Conference on Computer Science, Electrical and Electronic Engineering 2021 (ICCEE 2021), which was conducted through a virtual presentation on 18 August 2021. We would like to thank the colleagues and staff members at the institutions and organizations that served as partners for the international conference. Their support in organizing a successful conference has helped the editors to gather ideas and papers presented in this book. The editors are grateful to all the speakers who attended the conference and shared from their wealth of experience some exciting findings which have further propelled us to publish this book.

The editors also appreciate various people, including the production team at Springer, who helped and contributed to the creation of this book. We thank all the authors and contributors who presented at the conference and sent us their papers for peer-review. The editors would like to thank and appreciate the peer-reviewers for their suggestions, comments, efforts, and time spent to go over all the papers. The creation of this book has helped us to become a formidable team. The process has been enjoyable, challenging, inspiring, and more peaceful than we ever thought. We thank you all!

Prof. Dr. Zahriladha Zakaria
Dr. Seyed Sattar Emamian

Contents

Simulation of GSM Based Smart Energy Meter Presenting Electric Theft Detection and Prevention Mechanism by Using Arduino

Ateeb Hassan, Hadi Nabipour Afrouzi, Chua Hong Siang, Jubaer Ahmed, Kamyar Mehranzamir, and Chin-Leong Wooi

Abstract The design of a smart electricity meter relies on the technological needs of both utilities and customers, thus helping to face specific problems in the management and distribution of electricity. The most common issue is electricity theft that can be defined as the illegal use of electricity. Therefore, utility companies are on the lookout for ways to track and deter this fraud. This paper propose design for smart meter which presents a user-friendly design of smart energy meter that can detect the most common forms of theft in user premises in real-time. The design mainly consists of Arduino UNO, ACS712 current sensor, voltage sensor, and GSM module. Initially, some electrical parameters like voltage, current, and power are calculated. In addition, two current sensors are used that measures the electrical parameters of both phase and neutral lines. After that the electric theft is detected by comparing the values of these two sensors. Besides, a prevention mechanism is designed that prevent the electric theft by generating warning messages on LCD mounted on smart meter as well as sending the theft detection messages to utility through GSM and

A. Hassan · H. N. Afrouzi (✉) · C. H. Siang · J. Ahmed
Faculty of, Engineering Computing and Science, Swinburne University of Technology Sarawak, 93350 Kuching, Malaysia
e-mail: HAfrouzi@swinburne.edu.my

A. Hassan
e-mail: ahassan@swinburne.edu.my

C. H. Siang
e-mail: hschua@swinburne.edu.my

J. Ahmed
e-mail: jahmed@swinburne.edu.my

K. Mehranzamir
Department of Electrical and Electronic Engineering, Faculty of Science and Engineering, University of Nottingham Malaysia, Jalan Broga, 43500 Semenyih, Selangor, Malaysia
e-mail: kamyar.mehranzamir@nottingham.edu.my

C.-L. Wooi
Centre of Excellence for Renewable Energy (CERE), School of Electrical System Engineering, Pauh Putra Main Campus Universiti Malaysia Perlis, 02600 Arau, Perlis, Malaysia
e-mail: clwooi@unimap.edu.my

Z. Zakaria and S. S. Emamian (eds.), *Recent Advances in Electrical and Electronic Engineering and Computer Science*, Lecture Notes in Electrical Engineering 865, https://doi.org/10.1007/978-981-16-9781-4_1

turn off the supply for user with the help of relay. Furthermore, the proposed system is simulated in Proteus under normal condition and various theft scenarios and later verified via detailed simulation analysis. Finally, the system's ability is checked by implementing different theft techniques, and the figures in the paper indicate the warning messages associated with each theft scenario.

1 Introduction

Electricity is an extremely valuable resource. It is now an indispensable part of our everyday lives, and we cannot think of just one task without it. It has evolved into a crucial component for the survival of the greatest number of people and industries [1]. However, as the demand of electricity grows, so does the risk of electricity theft, which will continue until any steps are taken to identify and monitor it. Because of massive energy theft, energy distribution or consumption has recently become a hot topic of debate. A deliberate effort to steal a significant amount of energy by ensuring no or low energy logging in the metering system is referred to as theft in this case [2]. Electricity theft costs business a lot of money. Governments in some countries are losing revenue rather than making it. In certain cases, the government must provide incentives to the power industry to maintain fair energy prices. Due to financial losses, investments to boost existing capacity are limited and governments cannot meet the ever-growing electrical demand [3]. As a result, it is necessary to think along these lines and propose a solution to this obnoxious pattern.

Technical Losses (TL) and Non-Technical Losses (NTL) are the two types of losses in the transmission and distribution of electricity. TL occurs because of insufficient equipment sizing, dissipation in power lines, transformers, and other power station equipment. These losses are intrinsic and can be monitored. While meter tampering, meter malfunctioning, unauthorized connections, billing anomalies, fraud, pilferages, and incorrect meter readings are the most common NTL. Due to variety of factors, accurate measurement of these losses is difficult [4, 5]. Theft of electricity is the most common form of NTL. Bypassing the electricity meter, energy corruption of unregistered connections, interfering with the meter reading and direct hooking are all examples of this. It is liable for significant revenue losses as well as a drop in power efficiency and quality [4].

Various non-technical and technical approaches for detecting energy theft have been suggested in the past. Inspection of customers with suspicious load profiles is one example of a non-technical process. While periodic inspections can significantly reduce fraud, such a measure necessitates a significant amount of manpower and labor. In most situations, such efforts fail due to the staff's dishonesty. On the other hand, use of central observer meter at the secondary terminals of distribution transformer, harmonic generators, genetic support vector machines, external learning machines and the power line impedance techniques are some of the technical methods for detecting theft. However, these technical approaches can only be successfully implemented if adequate communication between the central control station and the required test points is maintained [3].

1.1 *Electricity Theft Methods, Detection and Protection*

Electricity failure and service interruption are usually the responsibility of electrical power utilities. End users of electricity, on the other hand, are similarly responsible for unusual power outages and shortages. Consumers make illegal and dishonest connections to avoid paying high energy bills. The power companies fail to prepare efficient and precise load shedding because of this unlawful use of electrical power. On the other hand, due to overburdening, this situation can cause electrical wire damage or increase distribution transformer faults. Consequently, power outage periods can be lengthened [6].

There are several ways to attempt the electricity theft, but phase line shorting, neutral line disconnection, tampering and whole meter bypassing by disconnection both phase line and neutral line are the popular ones. The Fig. 1 illustrates these theft methods so that it can be clearly understandable that how these theft attempts are made. Moreover, to detect these theft attempts two current sensors can be used, one on phase line and second on neutral line and their outputs are given to microcontroller. When someone tries to attempt the phase line shorting or neutral disconnection, microcontroller will detect a significant difference by comparing values of both sensors and then display warning message and communicate it with the utility [7].

To detect whole meter bypass a voltage sensor is used which will detect no voltages if both phase and neutral are disconnected. Then microcontroller will send power status signal to utility and after receiving confirmation about power availability it will generate a warning of whole meter bypassing [8]. Furthermore, meters can be tampered to display low or no energy consumption. Disc interruption, interfacing timing control unit used for two rate tariffs and reversing the meter are major types

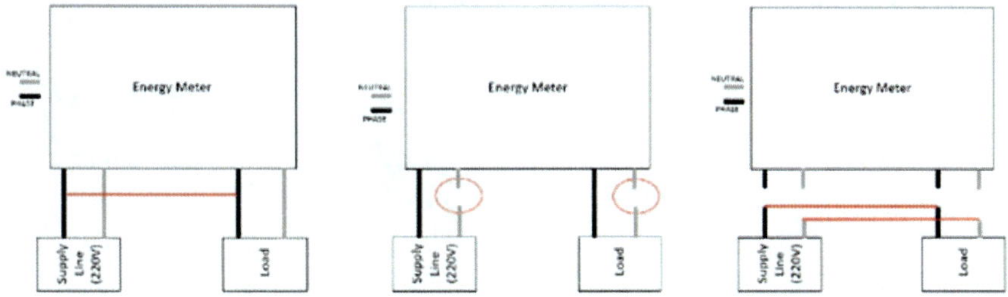

(a) (b) (c)

Fig. 1 Techniques used for implementing the electricity theft: **a** Phase shorting, **b** Neutral disconnection, **c** Whole meter bypassing of tampering. To detect tampering level switches can be used in meter boxes, when an unauthorized person will open the meter a tamper warning will be generated [2].

Some customers may try to connect an illegal or unregistered load to an existing load to obtain power for that illegal load. This type of theft is impossible to detect and cannot be tracked with a single household meter. Furthermore, accuracy of the electronics meters is tempered by using radio frequency devices. Electricity can be stolen in a variety of forms and with only a single household meter, these types of thefts are difficult to detect and avoid. The proposed method for detecting such unaccounted theft employs an observer meter. The observer meter measures the total energy usage of several households at fixed time intervals [9].

An observer meter is installed outside of an apartment to measure the energy consumption of all flats in that apartment. The measured values by household meter are then transmitted to the observer meter through SMS. After having two values central meter compares them and if there is a significant difference between these two values then it will detect it as a theft. The central meter will distinguish households where theft has occurred based on zero or low energy consumption since it has access to the readings of individual meter [10]. There are also other technical ways like harmonics generator, genetic support vector machines, external learning machines and the power line impedance techniques, but these methods can only be successfully applied if the central control station and the necessary test nodes can communicate effectively [3].

2 Design of Proposed System

A smart meter system is modelled, simulated, and evaluated that explains and assess the previously mentioned abilities to detect and defend against electricity theft. In Fig. 2a, the block diagram depicts the design structure of the smart energy meter system used in the simulation. Each line style in this diagram represents a distinct line form. The 220 V phase line is depicted in black (bold), while the neutral line is depicted in grey (thick). Finally, control signals are sent via the black (normal)

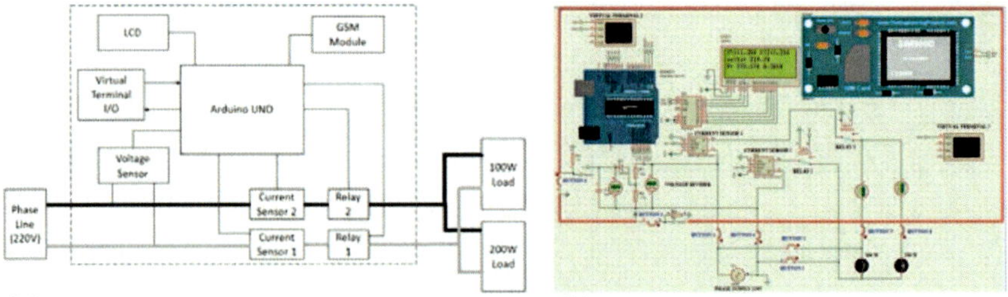

(a) (b)

Fig. 2 Diagrams of proposed system: **a** Block diagram, **b** Circuit diagram Proteus lines. Arduino is used as a microcontroller in this design. This designed model is used to simulate the smart energy meter system and to test various normal and theft operating scenarios. Meanwhile, the virtual circuit shown in Fig. 2b is created using the Proteus simulation software.

Proteus is a virtual system modelling (VSM) environment that allows you to simulate microcontroller code as well as electrical circuits in an effective way. For program the circuit, the code is written and compiled by using Arduino IDE after that the HEX file is used to simulate the circuit in Proteus.

The design of this system is mainly consisting of Arduino UNO, GSM module, ACS712 current sensor and voltage divider circuit. Arduino is used as a microcontroller and GSM module is used to send and received the messages from the utility side. Moreover, voltage sensor and two ACS712 current sensors are used to measure the voltages and current coming from the main line. Current sensor 1 is used to measure current of phase line and current sensor 2 is used to measure current of neutral line. Two relays 1 and 2 are used to connect phase and neutral line as shown in Fig. 2b. The voltage divider circuit has been used as a voltage sensor in the design. Basically, voltage divider scales down the voltages to level readable by Arduino and it is achieved with the help of two resistances, diode, and capacitor. The values of R1 and R2 are 100 k and 4.7 k, respectively. As only the positive cycle of the sine wave is required to measure the RMS voltages. Therefore, a diode is used which helps to cut off the negative cycle of the sine wave and only allow the positive cycle to pass. After that, a capacitor of 100uF is used to filter the incoming sine wave to reduce the noise level.

Two types of loads are used in this design where power rating of one bulb is 100 W and second bulb is 200 W that can be turned on or off with the help of button 7 and 8. The remaining buttons (1, 2, 3, 4, 5, 6) are used to test the circuit in different scenarios. LCD is used to display the readings and different messages to the user, and it is interfaced with the Arduino by using I2C protocol. To protect the system from tampering a red bold line is used to represent the case or box for meter and button 6 is used to simulate the tampering scenario. Besides, as there is no option in Proteus to receive the message through GSM module so two virtual terminals are used in the design to simulate the project. Virtual terminal 1 is used to send messages from utility side whereas virtual terminal 2 is used to display the messages that are sent to utility through GSM module. Finally, the proposed design includes a feature that allows the utility to remotely connect or disconnect the load if the customer fails to pay their bills.

3 Simulation Results of Proposed System

This section will present the simulation test results of proposed system under normal operating condition and different theft operation scenarios.

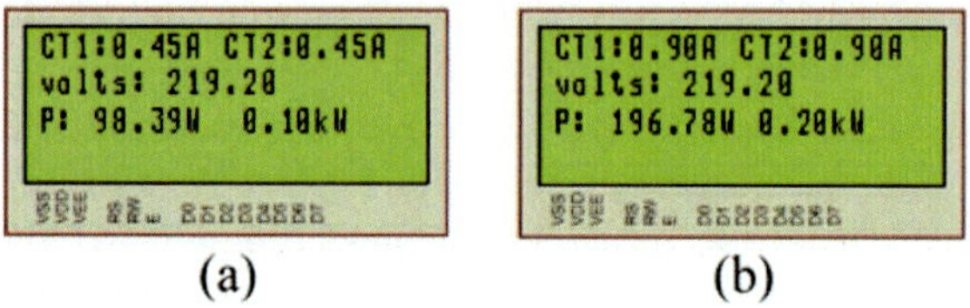

Fig. 3 Measured values of two different loads: **a** Current = 0.45A (100 W), **b** Current = 0.90A (200 W)

3.1 Results for Normal Operating Condition

The operation mode is called normal when values of CT1 and CT2 are same. Two different loads are used to check the accuracy of system. To turn on the 100 W load button 7 will be closed while button 8 will remain open. To turn on the 200 W load button 7 will remain open while button 8 will be closed. Figure 3 shows the values of current measured by the system for these two loads. The values of current for these two types of loads are verified by using ammeter available in Proteus. These values are also verified theoretically by using I = P / V equation.

3.2 Results for Phase Line Shorting Theft Method

If button 1 shown in Fig. 2b is pressed, then CT1 and relay 1 will be bypassed resulting the phase line shorting theft technique. Therefore, CT1 will measure less value then the CT2 and a current miss match case will be detected by Arduino so it will send the SMS to the Utility and displays the warning message along with a fine amount of 100RM on the LCD as well. The messages are shown in Fig. 4. As phase line is bypassed or shortened so user can use the loads by just using the neutral line coming from the energy meter. Therefore, cutting the neutral line by turning on the relay 2 will disconnect the supply for user.

When this theft method is applied proposed system will display warning message and ask user that if he wants to turn on the load then he must remove the bypassing wire from the phase line and must press the button 5. When user remove the bypass wire and press the button then Arduino will check the status of CT1 and again compare it with CT2. If there is no difference, then it will turn off the relay 2 and indicate that bypassing wire is removed. But if user did not remove the bypass wire and press the button 5 then it will indicate that bypassing wire is still not removed. That is

Fig. 4 Message during phase line shorting: **a** Warning message during phase line shorting, **b** Message for the user to take action

Fig. 5 Message during phase line shorting: **a** Warning message during neutral line disconnection, **b** Message for the user to act

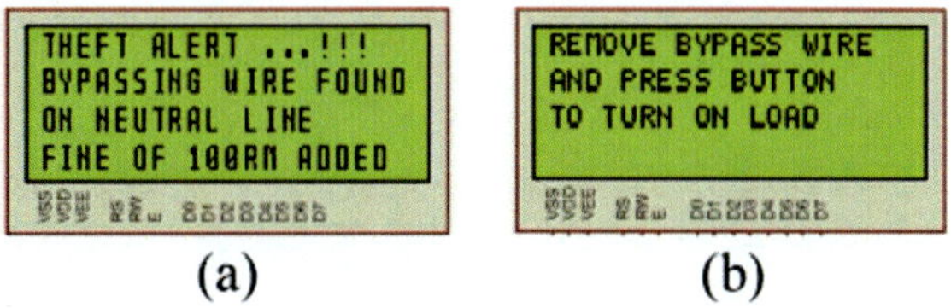

how the proposed system will detect and prevent the phase line shorting technique in real-time.

3.3 Results for Neutral Line Disconnection Theft Method

To simulate the neutral theft method, buttons 4 and 2 shown in Fig. 2b are used. When the button 4 is left opened while button 2 is pressed. In this case, CT2 and relay 2 will be bypassed. Therefore, the CT2 will measure less or zero current than CT1 and again a miss match scenario will be detected by Arduino. So, it will display the warning message along with fined amount of RM100 and send SMS to the utility about the theft detection along with the meter ID. The messages are shown in Fig. 5. In neutral theft detection case, the proposed system will turn on the relay 1 to restrict the user from using the load that he/she wants to use illegally.

Similarly phase line shorting method, when this theft method is applied the proposed system will display warning message and ask user that if he wants to turn on the load then he must remove the bypassing wire from the neutral line and must press the button 5. When user remove the bypass wire and press the button then Arduino will check the status of CT2 and again compare it with CT1. If there is no difference, then it will turn off the relay 1 and indicate that bypassing wire is removed. But if user did not remove the bypass wire and press the button 5 then it will indicate that bypassing wire is still not removed. That is how the proposed system will detect and prevent the neutral line shorting technique in real-time.

3.4 Results for Whole Meter Bypassing Theft Method

For the simulation and testing of whole meter bypassing, button 3 and button 4 shown in Fig. 2b are kept opened while button 1 and button 2 shown in Fig. 2b are pressed. In this case, the Arduino will detect no supply and record no value of charge being consumed. The voltage sensor will detect no voltages and this signal is captured by the microcontroller. To detect this type of theft, system will request the power status of the area by sending an SMS to utility. When utility confirms that electricity is available in that area the system will send an SMS to the utility about the theft detection along with meter ID. It will also ask to add fine of RM100 and send the

Fig. 6 Whole meter bypass warning message

Fig. 7 Warning message in meter tampering scenario

inspection team to the location. In addition, it will turn on the relays 1 and 2 shown in Fig. 2b and display the warning message on LCD that is shown in Fig. 6.

After restoring the phase and neutral line back by pressing button 3 and 4 and opening the button 1 and 2 the inspection team can send "resume" command to the proposed system to turn on the load by turning off relay 1 and 2. On the other hand, if utility confirm that electricity is not available in that area then proposed system will display "no electricity in the area" message on LCD and will continue to display until the electricity is provided back by utility.

3.5 Results for Meter Tampering

Users can sometime try to tamper the meter. Therefore, proposed system has a protection mechanism in case of meter tampering. Thus, meter tampering is detected when button 6 shown in Fig. 2b is kept opened. After detecting tamper, it will send a warning SMS along with the fine amount to utility and ask to send the inspection team to the location. Moreover, a warning message will also be displayed on the LCD which is shown in Fig. 7 and proposed system will turn on the relays to disconnect the supply for the faulty user. After checking and verifying, the inspection team will send the "close" command to the proposed system and after that the relays will be turned off and supply for the user will be continued.

Moreover, if an authorized person from utility wants to open the meter box, then he will send the "open" command and the proposed system will display a message on LCD that an authorize access is given to open the meter. Lastly, to close the meter box authorized person can send "close" command.

4 Conclusion

This paper is focused on the advancement of electrical energy metering systems and their potential to improve power consumption efficiency by detecting power theft and reducing unmeasurable losses. Moreover, a user-friendly theft detection system is presented that allows users to turn on their loads without the utility intervention in certain theft scenarios. Initially, some power theft methods have been presented to know about the techniques that users implement. Furthermore, a smart energy meter system capable of detecting power theft in real-time is simulated in Proteus, responsible for detecting the power theft and reporting the theft scenario to the utility. In addition, the working and testing of the proposed system are done in normal and theft operating conditions. The overall achieved specifications are that under standard power metering conditions, the device could reliably and precisely calculate the critical parameters like voltages, load current, and consumed power in W and kW. Shorting the phase line, disconnecting the neutral line, and bypassing the entire meter were all clearly demonstrated in the simulation model. Later, the ability of the designed system to accurately detect the various electricity theft methods by displaying a proper message and communication with the utility is demonstrated by simulation results in each case.

Finally, the use of machine learning algorithms, either by supervised learning or un-supervised learning is suggested as a possible future direction to detect the electric theft more precisely.

References

1. Shahid MB, Shahid MO, Tariq et al. H (2019) Design and development of an efficient power theft detection and prevention system through consumer load profiling. In: 2019 international conference on electrical, communication, and computer engineering (ICECCE), 24–25 July 2019. IEEE, pp 1–6. https://doi.org/10.1109/ICECCE47252.2019.8940644
2. S. Kumar.G, (2013) Power theft detection. Int J Technol Res Eng 4(8):2017
3. Mohammad N, Barua A, Arafat MA (2013) A smart prepaid energy metering system to control electricity theft. In: 2013 international conference on power, energy and control (ICPEC), 6–8 Feb. 2013. IEEE, pp 562–565. https://doi.org/10.1109/ICPEC.2013.6527721
4. Khan ZA, Adil M, Javaid N et al. (2020) Electricity theft detection using supervised learning techniques on smart meter data. Sustainability 12(19). https://doi.org/10.3390/su12198023
5. Hashmi MU, Priolkar JG (2015) Anti-theft energy metering for smart electrical distribution system. In: 2015 international conference on industrial instrumentation and control (ICIC), 28–30 May 2015. IEEE, pp 1424–1428. https://doi.org/10.1109/IIC.2015.7150972
6. Hussain Z, Memon S, Shah R et al (2016) Methods and techniques of electricity thieving in Pakistan. J Power Energy Eng 4(09):1–10. https://doi.org/10.4236/jpee.2016.49001
7. Ellenki SK, Srikanth Reddy G, Srikanth C (2014) An advanced smart energy metering system for developing countries. Int J Sci Res Educ 2(1):242–258
8. Dineshkumar K, Ramanathan P, Ramasamy S (2015) Development of AMR processor based electricity theft control system using GSM network. In: 2015 international conference on circuits, power and computing technologies [ICCPCT-2015], 19–20 March 2015. IEEE, https://doi.org/10.1109/ICCPCT.2015.7159401
9. Depuru SSSR, Wang L, Devabhaktuni V et al. (2010) Measures and setbacks for controlling electricity theft. In: North American power symposium 2010, 26–28 Sept. 2010. IEEE. https://doi.org/10.1109/NAPS.2010.5619966
10. Jaiswal VK, Singh HK, Singh K (2020) Arduino gsm based power theft detection and energy metering system. In: 2020 5th international conference on communication and electronics systems (ICCES), 10–12 June 2020. IEEE, pp 448–452. https://doi.org/10.1109/ICCES48766.2020.9138085

Model-Based Testing of Access Control Requirement in Multi-tenant Application: An Extensive Life Cycle

Gunavathi Duraisamy, Abdul Azim Abd Ghani, Hazura Zulzalil, and Azizol Abdullah

Abstract Model-based testing is a substantial approach that is based on and involving models. It is well known for achieving test coverage and for generating and executing test cases automatically. The main and core activity of model-based testing is the modelling activity. Models are used to explain and represent the behaviour of the product. The requirement will be translated into models and those models will be used in generating the test cases automatically in model-based testing. The modelling activity clarifies test requirement and contribute to automatic generation and execution of large test suites with tool support. Model-based testing provides numerous advantages in testing; mainly on achieving appropriate test coverage and reduce manual effort and time spent since it can execute the test cases automatically. Despite all the benefit it can provide, there are some drawbacks too. However, the challenges faced can overcome by implementing proper test modelling activity and by adopting a well-versed test management activity. Thus, in this paper, we have studied the existing testing process, the taxonomy of model-based testing and model-based testing steps and proposed an extensive lifecycle for model-based testing with detail activities and steps to be carried out in order to test and validate access control requirement for multi-tenant application.

G. Duraisamy (✉) · A. A. A. Ghani · H. Zulzalil
Faculty of Computer Science and Information Technology, Department of Software Engineering and Information System, Universiti Putra Malaysia, Serdang, Selangor, Malaysia
e-mail: gunavathi@utar.edu.my; gs43947@student.upm.edu.my

A. Abdullah
Faculty of Computer Science and Information Technology, Department of Communication Technology and Networking, Universiti Putra Malaysia, Serdang, Selangor, Malaysia

G. Duraisamy
Lee Kong Chian Faculty of Engineering and Science, Department of Internet Engineering and Computer Science, Universiti Tunku Abdul Rahman, Sungai Long Campus, Kajang, Selangor, Malaysia

Z. Zakaria and S. S. Emamian (eds.), *Recent Advances in Electrical and Electronic Engineering and Computer Science*, Lecture Notes in Electrical Engineering 865, https://doi.org/10.1007/978-981-16-9781-4_2

1 Introduction

Software testing is essential in the software development life cycle, even though it is complex and resource consuming activity. IEEE defines testing as an "activity in which a system or component is executed under specified conditions, the results are observed or recorded, and an evaluation is made of some aspect of the system or component" [7]. Testing always referred as an activity of finding faults or failure of the system under test (SUT). The main purpose of testing is to detect failure, find faults, discover defects and to check if the SUT is working as per intended and specified requirement. Also, testing is to ensure if the system is not behaving the way it is not supposed to behave. Releasing a product with failure and bugs will certainly bring bad impressions and a bad review from customers. Therefore, to gain customers' trust, it is the responsibility of the service provider to make sure that the product/application is well tested before release. Spending too much time and effort on testing alone is also not realistic and practical. The intensity of testing depends very much on the test techniques that are used and the test coverage that must be achieved. Test coverage serves as a test exit criterion [4]. Model-based testing (MBT) is well known for determining and achieving test coverage that necessary for the SUT [8, 10, 15, 18] and MBT is also known for generating and executing test cases automatically [8, 18].

Cloud computing and its applications have received significant attention among users in recent years. Although cloud provides new opportunities in a business perspective, it has also given a big impact on software testing and maintenance [3]. Testing security is one of the issue and challenges faces in cloud testing among the other issues. Data security and privacy of multi-tenant application has been a top concern of cloud users. In a multi-tenant environment, a security breach can result in the exposure of data to other, possibly competitive, tenants. This makes security issues such as data protection [5] very important at the application level instead of physical level. The service provider of SaaS applications should give more attention to this and the applications must be intelligent enough to segregate data of different tenant [11].

In order to avoid unauthorized access and to protect organizations assets, access control is one and the pioneer fundamental requirement of cloud computing [19]. Access control models are formal representation of access control policy, and the access control policies are the protection concepts of the application. Access control policies are high-level requirements that specify how access is managed and who, under what circumstances, may access what information [6]. Access control policies are also known as protection concepts. The security policies enforced by the access control system are represented by access control models (ACM). ACM are also useful to prove the system's limitation. In general, AC models are defined as a formalized computing algorithm, a well-recognized formal concept and as formally defined properties [6]. There are many access control models that are available in the industry. Role Based Access Control (RBAC) model is one of the earliest introduced model and known as the traditional model. The RBAC formal model was provided by David

Ferraiolo and Rick Kuhn in 1992 [12] and being widely used since then. As time evolves, the RBAC model being extended and updated as per the organization's needs and requirement. There are six basic data elements in RBAC: users, role, objects, operation, permission, and session [17]. The main concept of RBAC is 'role' which connects users and the operation of objects. RBAC can be used and applied in cloud computing with some modification and extension.

This paper presents an extensive life cycle for model-based testing approach to automatically generate test for access control rules and contracts of the associated activities based on access control requirements. Existing MBT is well known for functional testing and not much on testing security policies such as access control policies. Dianxiang Xu in his multiple works has implemented and proposed a model-based test generation technique using predicate and transition nets to test access control policy [18]. However, the work is not on multi-tenant environment. Thus, this paper will focus on the model-based testing that test access control policy for multi-tenant SaaS application which has been a challenge in recent years. In this paper we have studied the existing model-based testing process, taxonomy, and the modelling activities, then we extend and modify it to fit for model-based testing for access control of SaaS application in multi-tenant environment.

This paper is organized as in the section II will discuss about the existing model-based testing. Section III will discuss the different types of testing processes presented by ISTQB [4, 10], Utting et al. [15, 16], Zander et al. [20] and Dias-Neto and Travassos [1]. The section IV will talk about the proposed model-based testing life cycle for multi-tenant environment, arranged by the testing process and step-by-step explanation. Since this is an-ongoing research, the final section will discuss about the next action plans and future work.

2 Model Based Testing

According to ISTQB terms and definitions, model-based testing is a testing based on or involving models. The main idea of MBT is "to formalize and automate as many activities related to test case specification as possible and, thus, to increase both the efficiency and effectiveness of testing" [10]. In MBT, writing long huge pages of test case specifications are replaced by drawing a model (called MBT model) and tests are generated by tool support. Some MBT practitioners reuse models from UML diagram, software design and so on to generate the test cases. Some generate models from scratch referring to requirement specification or from access control policies depending on the test scope and coverage. These modelling approaches usually combine graphical and textual representations.

Model are used to explain and represent the behaviour of the product to give more understanding and to offer a reusable framework in the development of the product. Thus, selecting a model for MBT is an essential task and defining a clear modelling specification is important to generate useful test cases. Control-flow and data-flow strategy may use for test coverage. Dias-Neto and Travassos in his paper has addressed that dataflow-based strategy was used until 1999 and control-flow has become common after that [1].

MBT is a substantial approach and provide benefit not only for the testing activity but also for the whole product development process. The modelling process clarify test requirements [18] and since it is verified internally for completeness and consistency [9], it improves the understanding of the SUT. Also, this helps to improve the communication among developers and testers and, also in managing and mastering complexity of the test [8, 10]. The second main contribution of MBT is automatic generation and execution of large test suites easily and MBT is also supported by tool that significantly reduce time and effort spend on manual execution. The third benefit is that MBT models enables testers to visualize test and determine necessary test coverage. Besides that, model-based test models also cover most of the specified test objective, test levels and test types [10]. Since this increase the number and diversity of test cases, it also improves fault detection capabilities.

Despite all the advantages that MBT could provide, there are pitfalls in MBT too. The major challenge in MBT is that it demands skills and knowledge that testers require to understand and develop the test models. Thus, some substantial investment is needed to overcome this. Model that are generated and defined could have defects. This is common in all modelling activities. Therefore, it may require additional effort, time, skill, and other resources to validate the model. Another major and critical concern of MBT is that it can lead to test case explosion (also known as state-space explosion for state or other similar model). MBT is capable of generate huge number of test cases even from a small model. Thus, this again lead to an addition effort and time for model refinement, maintenance and checking and review. Likewise, all the other techniques and approaches, some of the drawbacks that MBT faces cannot be completely avoided and there is a need for a workaround to solve and improve it. Fortunately, many of the challenges can overcome with good skills and knowledge, by implementing proper test modelling activities and by adapting well versed test management activities (or software testing processes).

3 Model Based Testing Process

Software testing is a sequence of process to be carried out with the intention of identifying if the SUT is work properly and to find bugs. Test management is an activity to manage testing processes. There are many standards and research has proposed and outline testing processes and activity to be carried out to make sure that the developed product release with minimal issue. Some organizations are amending or modifying these testing processes to fit their organizational needs and convenience. One of the famous software testing board is International Software Testing Qualifications Board (ISTQB). It is a non-profit software testing qualification certification organization. They have presented the fundamental test process that should be carried out for software testing as general process and particularly for model-based testing. In this section, we will review and study software testing processes, activities and steps that has been proposed for general testing processes and for model specific testing.

Software Testing Life Cycle (STLC) is defined as a sequence of activities conducted to perform software testing [2, 13, 14, 21]. Figure 1 shows the list of activities to be carried out in the software testing life cycle. Testing usually starts after or during the requirement phase (from Software Development Life Cycle). Once the requirements are firmed, testers need to review the requirement or the design (if it is ready) to plan for the testing, to define the test objective and to determine the resources needed for testing. Once the requirement has been reviewed, test plan containing all the details regarding the testing activities will be prepared. From there, testing team will then start to design test and test cases will be created in line with the test scope and test objectives. Before the test execution, the test environment need to be setup and this step can be done any time after the test plan is confirmed and resources are ready. After the test is executed, results of the execution will be collected, and reports will be sent to the management. These are the general activities that are being carried out by most of the organization in the world. In this study, we are trying to apply and adapt this life cycle for model-based testing with some modification and extension that needed. Thus, we will study the processes proposed by ISTQB [4, 10], Utting et al. [15, 16], Zander et al. [20] and Dias-Neto and Travassos [1] to see if we can adopt some of the best practiced processes or process that can bring benefit to the model-based testing that we aim to test and verify the access control requirement.

Figure 2 represent the overall test process as defined by ISTQB. It starts with test planning, followed by analysis and test design, then implementation of test and test execution, evaluation of test exit criteria and report the outcome and finally with the test closure activity.

The STLC and the test process presented by ISTQB are for general testing process that can be applied for all kind of testing platform. We have first reviewed these two processes to come out with the list of process that can be applied to model-based testing. STLC has 6 processes in total and ISTQB has 5. Each of the processes are studies in detail and we have tried to match the processes that are performing same kind of actions or those that trying to accomplish same purpose. The outcome is presented in the following Table 1.

We can see that the requirement review is not in ISTQB and test environment setup processes from STLC are included or merged within the implementation process of

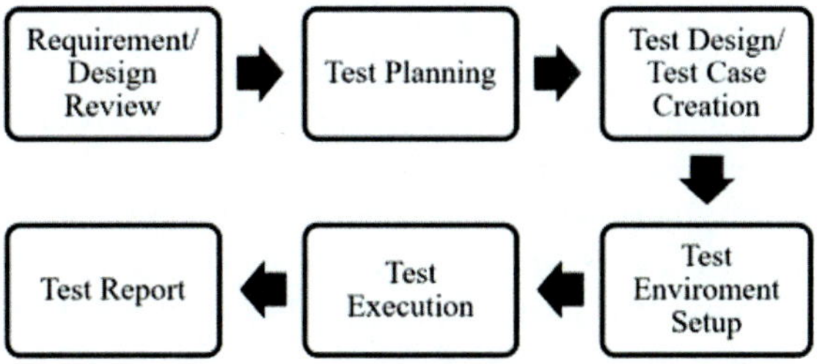

Fig. 1 Software testing life cycle (Adapted from [2, 13, 14, 21])

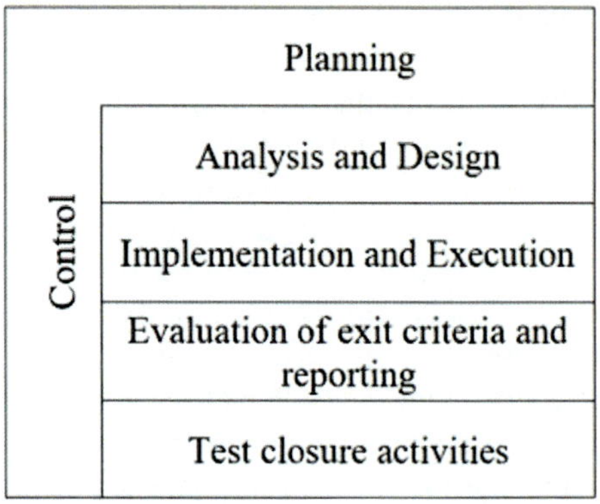

Fig. 2 ISTQB test process (Adapted from [4])

Table 1 General testing process

STLC	ISTQB
Requirement/design review	
Test planning	Planning
Test design	Analysis and design
Test case creation	Implementation
Test environment setup	
Test execution	Execution
	Evaluate exit criteria
Test report	Reporting
	Test closure

ISTQB. In STLC, these two are separate processes which has its own steps to be follow through. Separating these two processes is good in a way that it won't be missed out and test environment setup can be done any time after the resources are ready as per defined in the test plan. In STLC, after test execution, the next step will be preparing test report with all the findings and present to management and then decision will be made based on that. However, in ISTQB, test evaluation is an additional step and decision will be made based on the outcome of this process and documented in the test report. Test closure activity comprise of activities such as compiling the experience, lesson learnt and drawbacks from this whole testing process, archive test ware and hand over to the maintenance team. Test processes for model-based testing could be a bit different from the traditional testing process. It has some specific steps related to modelling activities since the test generation and execution process will be model based. ISTQB has presented a list of steps for model-based testing. The steps are simpler and mainly focuses on the modelling process, generation and execution of the test that are based on a specified model. Figure 3 shows the MBT specific activity as presented by ISTQB. Dias-Neto and Travassos in his paper has presented model-based testing activities as shown in the Fig. 4.

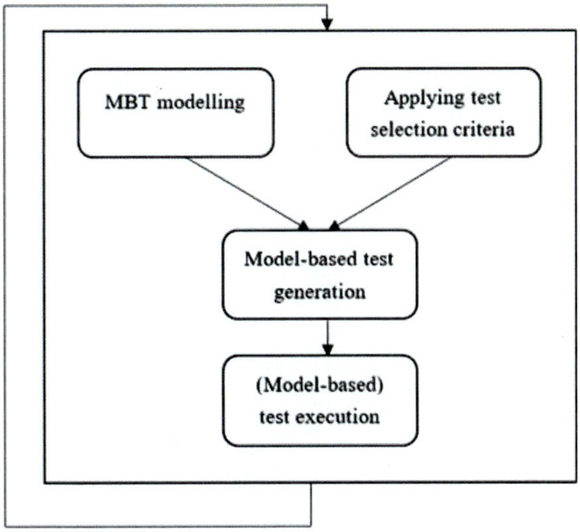

Fig. 3 Model-based testing steps by ISTQB (Adapted from [10])

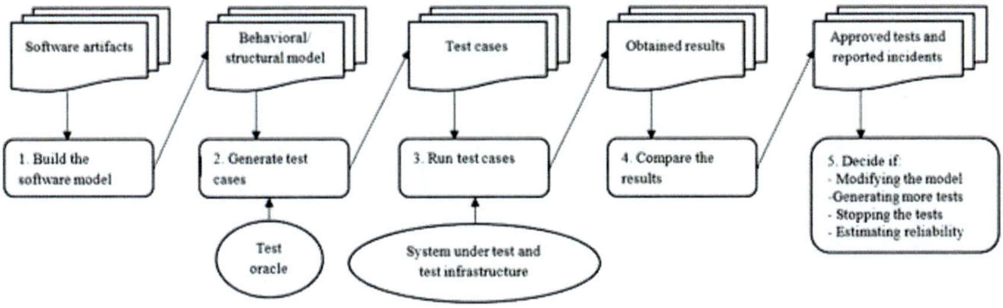

Fig. 4 MBT activities (Adapted from [1])

Utting et al. have presented a process and taxonomy for MBT and Zander et al. have complement test evaluation as an additional class. In the process, we can observe that the test model is based on the requirement and the requirement is also used for test selection criteria which then documented in test case specifications. Test cases are generated based on test case specification and test model. Utting et al.*'s* MBT process is shown in the following Fig. 5.

After review and studied the detail steps and actions of each processes proposed as above, we noticed that MBT has its main and important step which is the MBT modelling process. This process should be given more attention since it will be used and being core for generating and executing test which are model based. Similar to Table 1, we have studied each process and tried to match steps that are serving same purpose and present the outcome in Table 2.

As stated earlier, the modelling activity is the main process of model-based testing. The models are built from the requirement. Thus, requirement understanding is

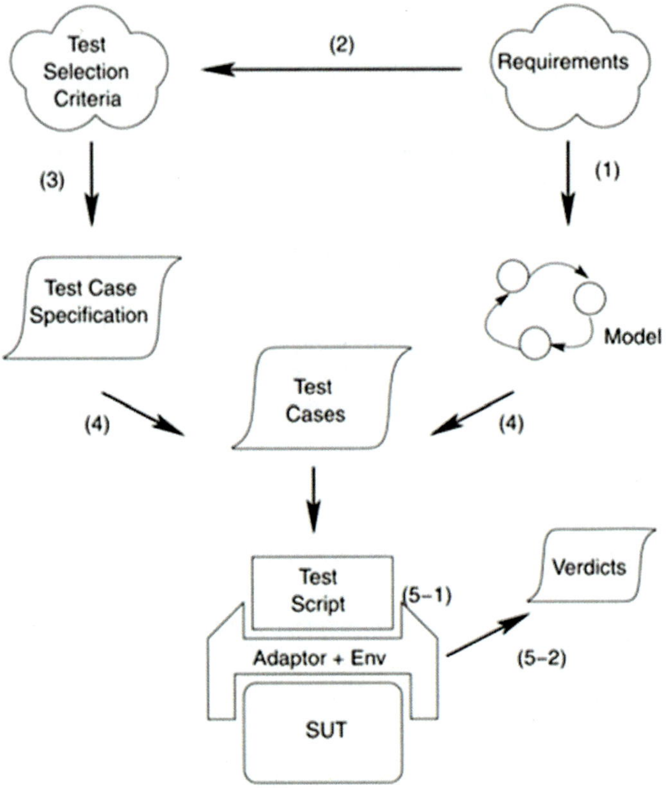

Fig. 5 MBT process by Utting et al. [16]

Table 2 Model-based testing steps

ISTQB	Utting et al. and Zander et al	Dias-Neto and Travassos
	Requirement	
	Test selection criteria	
MBT modelling test selection criteria	Model specification	Build the model
Apply test selection criteria	Test case specification	
MB test generation	Test generation: test case & test scripts	Generate test cases
MB test execution	Test execution: environment + adaptor + sut	Run test cases
	Test evaluation	Compare the results decide further actions

important and choosing a proper modelling approach is also crucial. Model-based testing can generate huge number of test even from a small model, thus it is important to define the test selection criteria and test coverage criteria. The test generation and execution can be done with the support of tool. There are many tools available for model-based testing in the industry. Selecting an appropriate tool is vital. All these decision for selecting proper tool, appropriate modelling approach and best test selection criteria should be aligned with the test objective, scope and coverage that aim to achieve. Therefore, it is important to have a test plan that will contain and document all this information which can be referred at any point of time during testing life cycle. After the test execution, not only the test evaluation is needed but also the test report and test closure activities are required to analyse and decide if there is any need to change the modelling approach based on the experience, test result and lesson learnt.

After analysing all the testing process (general and model-based), we can conclude that MBT steps should be extended with some modification. The following section will discuss the proposed MBT life cycle with detail process and steps. Since, this study is to test access control policies of multi-tenant SaaS application in public cloud the changes needed to cater this is also being highlighted.

4 Model Based Testing for Access Control

This section presents an extended life cycle of model-based testing process and steps to carry out in each of the defined process. Traditionally, the test models are generated based on the functional/non-functional requirement. This paper present model-based testing for access control which will use access control model for test modelling and to generate and execute test to verify and validate the security requirements (specifically the access control system) of the SUT. Figure 6 present an overview of proposed model-based testing life cycle for access control. Each process has its corresponding activity/s to be carried on.

Access Control Requirement Review: Requirement or design should be reviewed to get a clear understanding of the system under test. This will enable testers to get clear picture on how the system being build will behave and what are the expected features or behaviour of the system that customers are looking for. In this paper, as mentioned earlier our focus is on the access control security requirements of the SaaS multi-tenant application. Access control of the data security and privacy will be the main target of this study. Hence, testers should review the access control policy of the system under test first.

Step 1: Access Control Policy → Access Control Model: The access control policies will be transformed into a selected access control model that is suitable for the security requirement. In this research, we will be using Role Based Access Control Model with Ontology to represent the access control policies. The access control policies are written in XACML language.

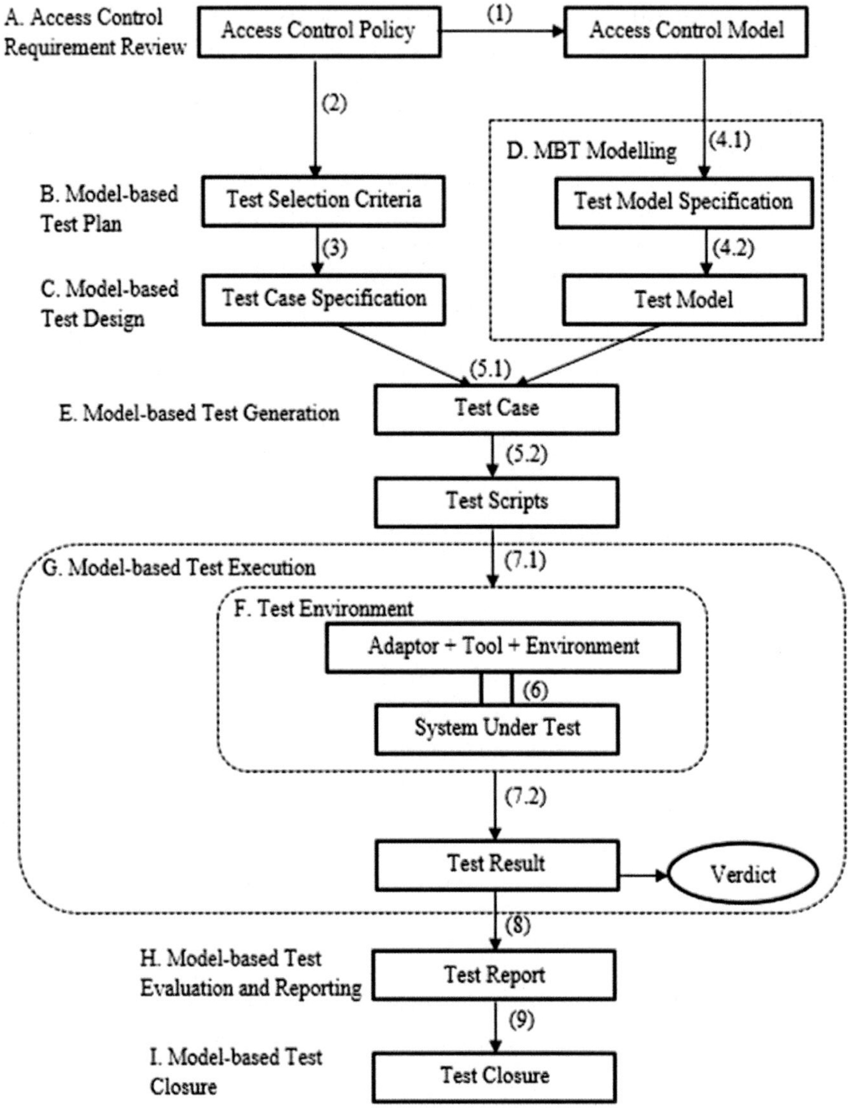

Fig. 6 MBT life cycle for access control

B. Model-based Test Plan: During this phase, the test policy and test strategies are determined based on the goal and objectives of the testing. Test selection criteria will specify the criteria that are suitable and appropriate for the objective of the test. It will also determine the scope and risk for the test cases that will be generated in later stage.

Step 2: Access Control Policy → Test Selection Criteria: From access control policy that already defined the access control requirements that need to be validated within the SUT, test selection criteria will be outlined. This will act as a guideline for test generation, to produce a test that will fulfil the objective of the testing. The coverage criteria of the test cases need to be defined here and the strategy and approaches to generate test cases are specified and explained in detail. All the detailed specification will be based on the test objective and requirement that has been set earlier.

C. Model-based Test Design: The objective of this phase is to transform the objective of the test into concrete test conditions and design.

Step 3: Test Selection Criteria → Test Case Specification: Test case specification is the high-level description of the test cases. Based on the test policy and objective, and from the test selection criteria, the test case specification will be produced. The objective of each test cases will be defined, and the input, procedure and expected output is stated in the specification. The scenarios to be covered, how and how often they will be tested will also be described. Test data that need and will be used in the testing need to define along with the input and output specification.

D. MBT Modelling: This phase will contain two main activities of MBT modelling. Those are first to define the test model specification and second is to build a test model based on the specification. Test model specification will be detailed out in for modelling process that are suitable and appropriate for the objective of the test. The test model specification also will determine the scope and risk for the test model.

Step 4.1: Access Control Model → Test Model Specification: As mentioned earlier, access control model is the formal representation of the access control policy. The access control requirement will be transformed into access control model. Each ACM has its own properties and characteristics. In this study, RBAC with ontology is used as the ACM, thus the test model specification is related and specified to align RBAC and ontology properties. The policy, scope and the structure of the model will be explained. Also, the coverage criteria need to be defined to make sure there is no test case explosion happened during test generation stage.

Step 4.2: Test Model Specification → Test Model: Test modelling process will take place during this stage and modelling of the test model will based on the specification drawn in the previous stage. There are many modelling behaviours available in the industry that being used by the model-based testing. It is important to choose the modelling behaviour that is appropriate and suitable for the objective and scope of the testing. Based on the related studies, we found that modelling behaviour that uses petri-net are proven to generate and build reliable test models for access control. This study is also will use petri-net in the modelling process to generate the test model.

E. Model-based Test Case Generation: This phase will focus on the generation of test cases. Since, model-based testing has advantage of generating test cases automatically, this study will aim to generate test automatically by define test modelling with tool support. The tool used will incorporate the test model specified in earlier stage and will transform the test model into executable test cases automatically. The tool used in this study also need to be associated with the modelling behaviour that has been chosen.

Step 5.1: Test Model + Test Case Specification → Test Case: Test cases are generated automatically from the test model and the test case specification that have been defined earlier. The test cases are generated in the aim to satisfy all the test case specification and in line with the test objectives.

Step 5.2: Test Case → Test Script: The test model can generate huge number of test cases, even with the defined test coverage in the test case specifications. Therefore, this step is needed to determine the test cases to be executed. During this stage, the test cases need to be prioritized and the appropriate test will be selected and included in the test scripts to be executed.

F. Test Environment Setup: The test environment needs to be setup beforehand and not necessary to wait till the test scripts are ready. This step can be an independent process which can be started once the environment requirement and resources are clearly defined in the test plan. The environment for SUT, for the testing tool and the adaptor that needed to connect both need to clearly be chosen and define before the test execution activity starts.

Step 6: Test Environment Setup: (Adaptor + Tool + Environment) + SUT: For this study, the system under test is a SaaS application in public cloud and it is in the multi-tenant environment. Thus, the proper environment to accommodate the SUT need to be setup. The test generation and execution process will be automatic, and this is done by support from the selected model-based testing tool. Therefore, the environment and resources needed for the testing tool need to be arranged. The environment for both should be at the same and must ensure that this will support both SUT and testing tool to work with each other without any discrepancies.

G. Model-based Test Execution: During this phase, the generated and selected test scripts will be executed. The test environment details, SUT version, identities of test tool used, and any test data used should be logged into test log. The outcome of this activity needs to be logged to compare the actual result with expected results.

Step 7.1: Test Execution: Test Scripts + Test Environment: The selected test scripts need to be executed in the defined test environment. This process is also automated by tool support and by environment that provide needed facility. The testing tool used will execute test cases and record the test verdicts automatically.

Step 7.2: Test Execution → Test Result: Once each of the test execution process is complete, the outcome of the test needs to be recorded as test results. The results then used to compare the actual output and the expected output of the test cases. The discrepancies are reported as incidents (defects/bugs) and this need to be rectified. After the rectification, the test activities are repeated for each discrepancy and test re-executed with regards to the rectification made.

H. Model-based Test Evaluation and Reporting: Test evaluation is an activity to compare the actual SUT output with the expected SUT behaviour based on test oracle.

Step 8: Test Result → Test Report: Based on the test result and test log that collected in the previous phase, test summary report is prepared during this stage. Test log is checked against the exit criteria set or defined in test plan and it is assessed against the defined test objectives. From the outcome, decision is made on whether more test is needed, or the exit criteria need to be change. All the outcome and decisions will be recorded in the test report document. The test report also will be used by the management to review the SUT's capability in achieving the intended requirement and the testing activity performance as whole.

I. Model-based Test Closure: This activity is to consolidate the lesson learnt, experience gained, and any risk or problem faces, and the solution or action taken to overcome it. This activity usually took place after the SUT is delivered to the customer.

Step 9: Test Report→ Test Closure: From the test report, the management and the testing team can measure the actual deliverable versus the planned deliverables of the whole testing process. This is to gauge which planned deliverables are successfully delivered and if the discrepancies that reported as incidents being solved and verified before release to customer. Then, the test wares are handed over to the maintenance team or being archived. Finally, the team need to evaluate how the whole process is went and analyse the experience gained and the lesson learnt to make use of them in future.

5 Conclusions and Future Work

Software testing is an important part in the software development. The application or product being build must be carefully and well tested before release to customers. The main aim of testing is to find faults and to make sure that the application is behaving as per requirement. Model-based testing is well known for achieving necessary test coverage and it can automatically generate and execute test following a proper modelling approach. Implementing a proper modelling activity and adapt a suitable test management activity can help to overcome the challengers faced by the model-based testing. This paper has reviewed testing process and life cycle that proposed by ISTQB and other researchers. As the result of the analysis, we found that there is a need to extend the proposed steps and activity to cover all the aspects that need to be consider in a testing phase. Then this paper has presented a model-based testing life cycle that comprises detail processes and steps to be carried out. This model-based testing life cycle is aimed to help testers to conduct a comprehensive testing on validating the access control requirement in multi-tenant environment. As mentioned earlier, this is an on-going research and the future work will be to implement the proposed model-based testing process to test access control requirement of e-health SaaS application which is hosted in multi-tenant environment. Each proposed step and activities will be followed, and test result and outcome will be recorded. The test results will then be evaluated using mutation analysis to measure if the proposed modelling approach is effective in fault detection and improve effectiveness by reduce manual effort of test generation and execution.

Acknowledgements This work was supported in part by FRGS under Grant Nos. 08-01-16-1848FR.

References

1. Dias-Neto AC, Travassos GH (2010) A picture from the model-based testing area: Concepts, techniques, and challenges. In: Advances in Computers, vol. 80, pp 45–120. Elsevier
2. Everett GD, McLeod R Jr (2007) Software testing: testing across the entire software development life cycle. Wiley
3. Gao J, Bai X, Tsai WT, Uehara T (2013) Saas testing on clouds-issues, challenges and needs. In: 2013 IEEE 7th international symposium on service oriented system engineering (SOSE). IEEE, pp 409–415
4. Graham D, Van Veenendaal E, Evans I (2008) Foundations of software testing: ISTQB certification. In: Cengage learning EMEA
5. Guo CJ, Sun W, Huang Y, Wang ZH, Gao B (2007) A framework for native multi-tenancy application development and management. In: The 9th IEEE international conference on e-commerce technology and the 4th IEEE international conference on enterprise computing, e-commerce, and E-services, 2007. CEC/EEE 2007. IEEE, pp 551–558
6. Hu VC, Kent KA (2012) Guidelines for access control system evaluation metrics. US Department of Commerce, National Institute of Standards and Technology
7. IEEE Standard Glossary of Software Engineering Terminology (1990) IEEE STD 610.12–1990. https://doi.org/10.1109/IEEESTD.1990.101064
8. Jorgensen PC (2017) The craft of model-based testing. CRC Press
9. Kiran M, Friesen A, Simons AJ, Schwach WK (2013) Model-based testing in cloud brokerage scenarios. In: International conference on service-oriented computing. Springer, Cham, pp 192–208
10. Kramer A, Legeard B (2016) Model-based testing essentials-guide to the ISTQB certified model-based tester: foundation level. Wiley
11. Subashini S, Kavitha V (2011) A survey on security issues in service delivery models of cloud computing. J Netw Comput Appl 34(1):1–11
12. Sandhu RS, Coyne EJ, Feinstein HL, Youman CE (1996) Role-based access control models. Computer. http://doi.org/https://doi.org/10.1109/2.485845
13. Tang J (2010) Towards automation in software test life cycle based on multi-agent. In: 2010 international conference on computational intelligence and software engineering (CiSE). IEEE, pp 1–4
14. Tekin O, Cetin GB (2012) Application test process in product life cycle. In: 2012 6th international conference on application of information and communication technologies (AICT). IEEE, pp. 1–6
15. Utting M, Legeard B (2010) Practical model-based testing: a tools approach. Morgan Kaufmann
16. Utting M, Pretschner A, Legeard B (2012) A taxonomy of model-based testing approaches. Softw Test, Verific Reliab 22(5):297–312
17. Wang W, Han J, Song M, Wang X (2011) The design of a trust and role based access control model in cloud computing. IEEE, pp 330–334. http://doi.org/https://doi.org/10.1109/ICPCA.2011.6106526
18. Xu D, Kent M, Thomas L, Mouelhi T, Le Traon Y (2015) Automated model-based testing of role-based access control using predicate/transition nets. IEEE Trans Comput 64(9):2490–2505
19. Younis YA, Kifayat K, Merabti M (2014) An access control model for cloud computing. J Inf Secur Appl 19(1):45–60
20. Zander J, Schieferdecker I, Mosterman PJ (2011) A taxonomy of model-based testing for embedded systems from multiple industry domains. In: Model-based testing for embedded systems, pp 3–22
21. Zhang T, Su Y, Wang J, Wang J (2017) A novel model for software development and testing in programmable logic. In: 2017 IEEE international conference on software quality, reliability and security (QRS). IEEE, pp 81–85

Survivable Biconnected Topology for Yemen's Optical Network

Omar Khaled Omar Baslaim, Farabi Iqbal, Sevia Mahdaliza Idrus, and Abu Sahmah Mohd Supa'at

Abstract Optical fiber networks are vital for providing important telecommunication services worldwide. Many developing countries are in the emerging phases of deploying their optical fiber networks, and need to consider the incurred capital expenditure cost, while ensuring the network robustness during the planning phase, deployment phase and upgrade phase of their network infrastructures. The network topology planning needs to meet desired specifications (e.g., connectivity and robustness), while lowering overall deployment costs. A network must be able to remain connected and provide services, even after the event of any node or fiber failures. In this paper, we propose an integer linear program formulation and a heuristic for finding the minimum cost biconnected network topology. We use the Yemeni topology as the case study and show prospective biconnected network topologies for it.

1 Introduction

Optical fiber has transformed the telecom landscape in the world in the last three decades, due to its overwhelming advantages over the other transmission mediums [1]. Optical fibers offer low loss connectivity with high bandwidth and long operational lifespans. Example of challenges for network providers in deploying the

O. K. O. Baslaim · F. Iqbal (✉) · S. M. Idrus · A. S. M. Supa'at
School of Electrical Engineering, Faculty of Engineering, Universiti Teknologi Malaysia, 81310 Johor Bahru, Malaysia
e-mail: alfarabi@utm.my

O. K. O. Baslaim
e-mail: omar1990@graduate.utm.my

S. M. Idrus
e-mail: sevia@utm.my

A. S. M. Supa'at
e-mail: abus@utm.my

Z. Zakaria and S. S. Emamian (eds.), *Recent Advances in Electrical and Electronic Engineering and Computer Science*, Lecture Notes in Electrical Engineering 865, https://doi.org/10.1007/978-981-16-9781-4_3

fiber infrastructure are the cost of the network component and equipment, and network robustness. Economic constraints lead to the importance of cost-saving measures (e.g., wavelength-division multiplexing [3, 4] and fiber co-deployment [5]) for network operators, though the network must be able to provide dependable services [2] to the ever-increasing rising number and needs of clients. Cost-saving of network connectivity can be achieved via the use of biconnected network topology with minimum deployment cost [2, 6, 7]. Robustness is the ability of a network to continue working well even in the events of network faults [8]. Fiber cuts, configuration mistakes, viruses/worms, cyber-attacks, terrorism, and natural calamities can all cause network failures [9]. Hence, analyzing the robustness of national optical networks aids in understanding and mitigating the impact of network failures [10]. The exact price of laying fiber differs by country and depends on many parameters, such as cable type, number of fiber mode, cost factors, installation techniques and strategies, rules and challenges that differs from country to country [11–14]. The minimum-cost network topology design for given node locations and traffic forecast must also balance network survivability and deployment cost, which may have conflicting objectives. Selecting a suitable topology for communication network would be beneficial to the long-term operation since the topology would have a large impact on the performance and characteristics of the network. Yemen still have lot of opportunities to improve its telecommunication infrastructure. The design capacity of submarine cables serving per capita in 2011 is 53 kbps, with two submarine cables connected to Yemen [15, 16]. Exclusive rights, acquisition of right of ways, implementation information requests, and ongoing civil wars need also be considered for fiber deployments.

2 Proposed Approach

A graph *G(N,E)* is a representation of network topology, where *N* represents a set of vertex and *E* represents a set of edges connecting two certain vertices in *N*. A graph *G* is connected if and only if every two vertices in the graph are connected by a path in *G* (shown in Fig. 1). Otherwise, *G* is disconnected. A path is a sequence of edges that lead from a vertex to another in the graph. The path length is the total

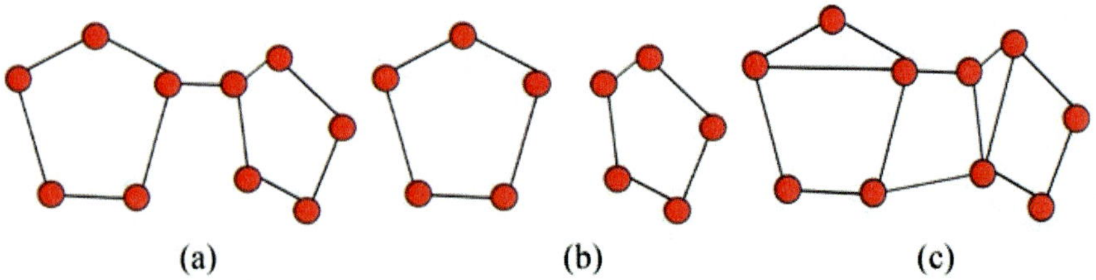

Fig. 1 **a** Connected graph, **b** disconnected graph, **c** biconnected graph

length of edges consisting the path itself. Edge length is the length of the edge that connects two vertices. A graph is k-vertex-connected if a graph is still connected after removal of *(k–1)* vertices. A graph is k-edge-connected the if graph still connected after removal of *(k–1)* edge. Biconnected graph, as shown in Fig. 1c ensures that there are at least two vertex-disjoint paths [17] for each connection, such that the graph is still connected after removal of any single vertex and minimizing the total network deployment cost.

In our first approach, we formulate an integer linear program (ILP) that returns the biconnected topology with minimum total edge length.

ILP Constants and Variables.

R set of possible vertex pairs.

A_{uv} is 1 if edge (u, v) exists in G; 0 else.

L_{uv} length of edge (u, v).

T_{rf} set of r number of entries containing all possible vertex pair $f = (n_1, n_2)$.

P_{ruv} is 1 if the primary path of entry $r \in R$ uses edge (u, v) in G; 0 else.

B_{ruv} is 1 if the backup path of entry $r \in R$ uses edge (u, v) in G; 0 else.

X_{uv} is 1 if there is the network operator needs edge (u, v); 0 else.

ILP Objective & Constraints.

Minimize

$$\sum_{u \in N} \sum_{v \in N} X_{uv} \times L_{uv} \tag{1}$$

Any connection entering an intermediate vertex should exit it.

$$\sum_{u \in N} P_{uk} = \sum_{v \in N} P_{kv} \forall r \in R, k \in N : k \neq T_{rn_1} \text{or} T_{rn_2} \tag{2}$$

$$\sum_{u \in N} B_{uk} = \sum_{v \in N} B_{kv} \forall r \in R, k \in N : k \neq T_{rn_1} \text{or} T_{rn_2} \tag{3}$$

Paths start from the source vertex and end at the destination vertex.

$$\sum_{u \in N} P_{ruT_{rn_2}} = \sum_{v \in N} P_{rT_{rn_1}v} = 1 \forall r \in R \tag{4}$$

$$\sum_{u \in N} P_{rT_{rn_2}u} = \sum_{v \in N} P_{rvT_{rn_1}} = 0 \forall r \in R \tag{5}$$

$$\sum_{u \in N} B_{ruT_{rn_2}} = \sum_{v \in N} B_{rT_{rn_1}v} = 1 \forall r \in R \tag{6}$$

$$\sum_{u \in N} B_{rT_{rn_2}u} = \sum_{v \in N} B_{rvT_{rn_1}} = 0 \forall r \in R \tag{7}$$

$$P_{ruv} + P_{rvu} + B_{ruv} + B_{rvu} \leq 1 \forall r \in R, u \in N, v \in N \tag{8}$$

$$\sum_{v \in N} P_{ruv} + \sum_{v \in N} B_{ruv} \leq 1 \forall r \in R, u \in N : u \neq T_{rn_1} \tag{9}$$

Paths are routed through viable edges.

$$P_{ruv} + B_{ruv} \leq A_{uv} \forall r \in R, u \in N, v \in N \tag{10}$$

$$P_{ruv} + B_{ruv} \leq X_{uv} \forall r \in R, u \in N, v \in N \tag{11}$$

Since our first approach can be time consuming for networks with high number of vertices, we also propose a two-phase heuristic approach that is faster, but at the expense of optimality for solving the problem. In the first phase, we use find a topology where each network vertex has at least two node degree (neighbouring vertices) and minimized total edge length. If the network is not biconnected, we augment the network topology by carefully choosing the best edge to be added into the topology, based on the vertex betweenness centrality [8, 18]. Betweenness centrality is proportional to the number of shortest paths that travel through the vertex. The transitional steps are shown in Fig. 2. Vertices will be separated into groups and the group with the highest betweenness centrality will be chosen and

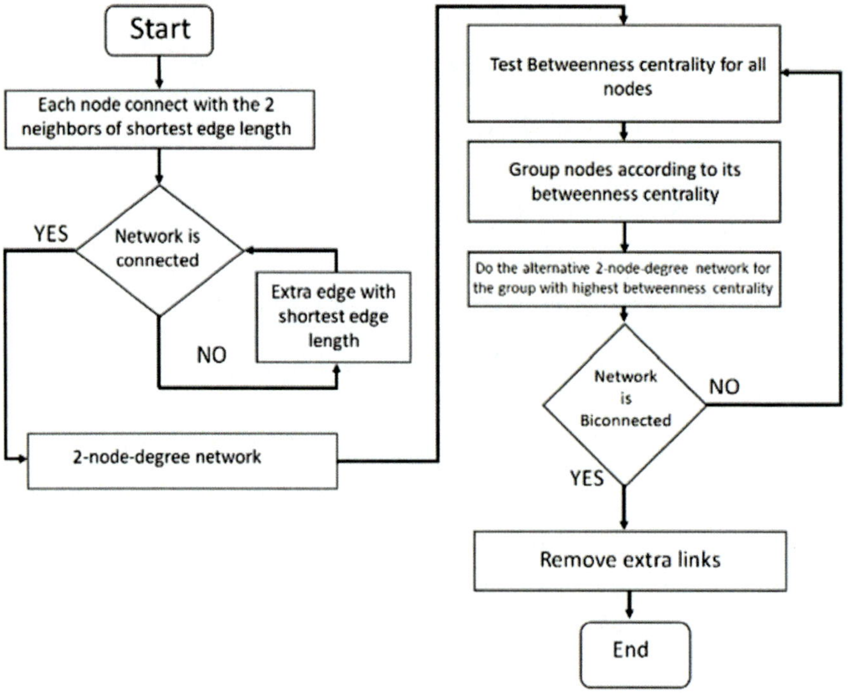

Fig. 2 Proposed two-phase approach

all corresponding connected vertices removed. Then, we rerun phase 1 to find the alternative suitable edges for improving the network connectivity. Once we have a biconnected topology, we will remove all redundant edges to get the solution.

3 Results and Discussion

There are 21 governorates for Yemen, and we assume them as prospective vertices for the Yemen optical network. Figure 3 shows the distribution of vertices in the center of the cities and possible edge placement between the vertices.

From the ILP formulation, we attain a biconnected ring topology with minimized total edge length, as shown in Fig. 4. Simulations were conducted on an Intel(R) Core i7-7700 3.6 GHz machine of 8 GB RAM memory. On the other hand, Fig. 5 shows the result from our heuristic. From phase 1, we attain the topology where each vertex is at least two node degree and the total edge length is minimized as shown by Fig. 5a. Since this topology does not fulfil the biconnectivity requirement, we add more targeted edges with phase 2 (based on groups of vertices that are sorted based on their betweenness centrality) as shown in Fig. 5a until biconnectivity is achieved. When the network is biconnected, we can reduce the number of edges to the result shown in Fig. 5b.

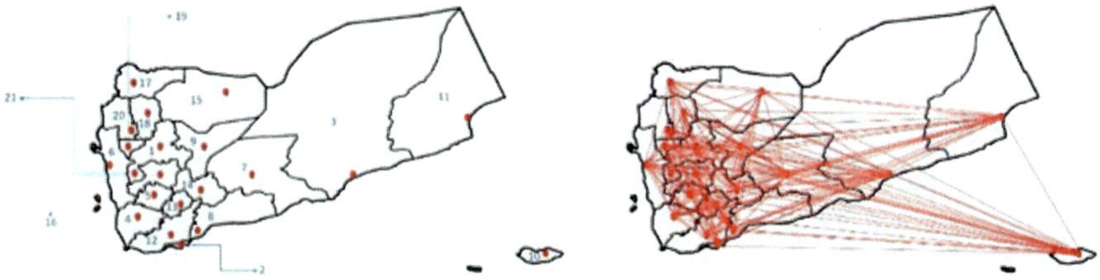

Fig. 3 Yemen governorates and the complete graph topology

Fig. 4 ILP output

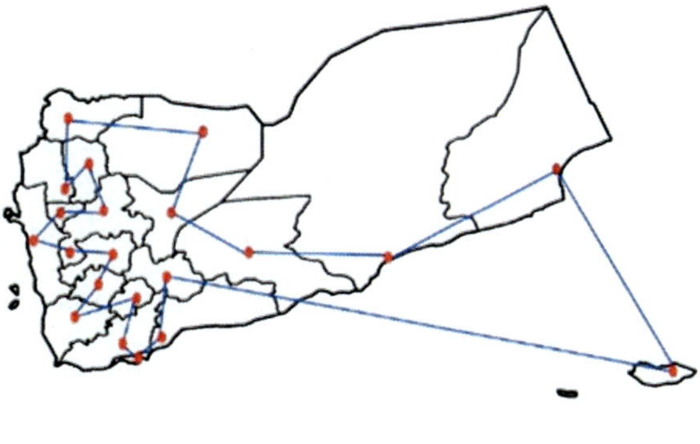

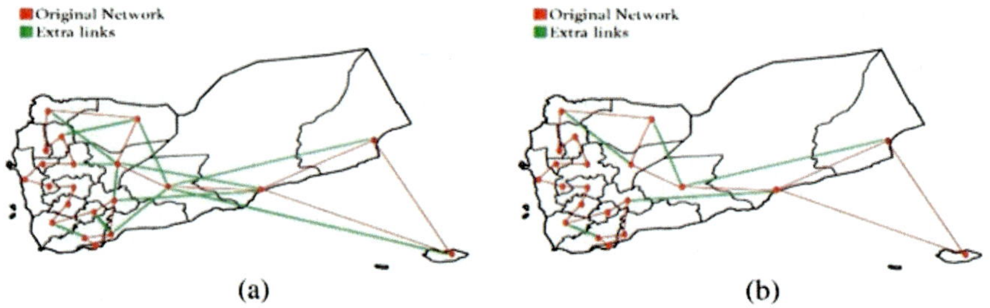

Fig. 5 **a** Phase 1 and 2 output. **b** Topology with five extra edges

4 Conclusion

In this paper, we have proposed an ILP and a heuristic for finding the minimum cost biconnected network topology, using the Yemeni topology as the case study. By reducing deployment expenses, while ensuring survivability, the network operators can have an initial min-cost survivable backbone network that can then further be augmented to suits their future needs. Possible future directions are geographical limitations, primary/backup path length limit, and nodal placement optimization.

Acknowledgements This work was supported by Ministry of Higher Education Malaysia through Universiti Teknologi Malaysia institutional grant vote 05G28 and 05G27.

References

1. Agrawal GP (2016) Optical communication: its history and recent progress. Springer
2. Scheffel M (2005) Optimal topology planning of optical networks with respect to overall design costs. Opt Switch Netw 2(4):239–248
3. Ismail MM, Othman MA, Zakaria Z et al (2013) EDFA-WDM optical network design system. Proc Eng 53:294–302
4. Singh S, Singh A, Kaler RS (2013) Performance evaluation of EDFA, RAMAN and SOA optical amplifier for WDM systems. Optik 124(2):95–101
5. ESCAP (2018) A study on cost-benefit analysis of fibre-optic co-deployment with the Asian highway connectivity, p 49
6. Morales FG, Paiva MH, Bustos-Jiménez JA (2018) Measuring and improving network robustness: a Chilean case study. IEEE Commun Lett 23(1):44–47
7. Zhao, R, Minge C, Schweigel M (2009) Enhanced survivable topology redesign of optical broadband networks with biconnectivity. Int J Adv Telecommun
8. Rueda DF, Calle E, Marzo JL (2017) Robustness comparison of 15 real telecommunication networks: structural and centrality measurements. J Netw Syst Manag 25(2):269–289
9. Ashraf MW, Idrus SM, Iqbal F et al (2018) Disaster-resilient optical network survivability: a comprehensive survey. MDPI Photon 5(4):35
10. Pavan C, de Lima LS, Paiva MHM et al (2015) How reliable are the real-world optical transport networks. J Opt Commun Netw 7(6):578–585
11. CTC (2009) Brief engineering assessment: cost estimate for building fiber optics to key anchor institutions Cost of Building Fiber to America's Anchors

12. Kamau GM (2015) A model for estimating Netw. infra. A Case for All-Fibre Netw, Costs
13. Nyarko-Boateng O, Xedagbui FEB et al (2020) Fiber optic deployment challenges and their management in a developing country: a tutorial and case study in Ghana. Eng Reports 2(2)
14. Hayford-Acquah T, Asante B (2017) Causes of fiber cut and the recommendation to solve the problem. IOSR J. Electron. Commun. Eng 12:46–64
15. Al-Madhagy T (2013) ICT policy in Yemen. University Utara Malaysia
16. Gelvanovska, N, Rogy M, Rossotto CM (2014) Infrastructure deployment and developing competition. Directions in development-information and communication technologies
17. van Adrichem NL, Iqbal F, Kuipers FA (2016) Computing backup forwarding rules in software-defined networks. arXiv preprint arXiv:1605.09350.
18. Zainiar NA, Iqbal F, Supa'at ASM, et al (2020) Robustness metrics for optical networks. Indonesian J Electr Eng Comput Sci 20(2):845–853

A Study on Electric Field Distribution in Polymeric Insulator Using Finite Element Method

Law Kim Yin, Hadi Nabipour Afrouzi, Ateeb Hassan, Jubaer Ahmed, Kamyar Mehranzamir, and Saeed Vahabi Mashak

Abstract Manufacturing failures of insulators, such as the existence of cavities due to air bubbles and impurities during polymer injection can cause electric field distortion to occur where the air ionizes and fills the internal voids leading to the occurrence of partial discharges. This paper analyzes the maximum electric field and region where the field exists are studied with the model of the insulator's configuration and analyzed using Finite Element Method (FEM). Contamination will also be simulated to compare the severity between the cavity and pollution. It was observed that pollution is more severe in polymeric insulators when compared to the presence of cavities. This research is significant as early PD detection is necessary to prevent faults in the insulation systems.

1 Introduction

High voltage insulators must withstand electrical and mechanical stresses as it could cause corona discharges and degradation of the insulator which could ultimately lead to an insulation flashover, causing faults in the power system [1–3]. This is due to the electric field distribution along the length and surface of the insulators which affects electrical performances. The higher the conductivity of the pollutant, the higher the electric field strength which increases the possibility of flashover to occur [4]. The position of the cavity in an insulator is also critical to determine its electric field

L. K. Yin · H. N. Afrouzi (✉) · A. Hassan · J. Ahmed
Faculty of Engineering Computing and Science, Swinburne University of Technology Sarawak, 93350 Kuching, Malaysia
e-mail: HAfrouzi@swinburne.edu.my

K. Mehranzamir
Faculty of Science and Engineering, Department of Electrical and Electronic Engineering, University of Nottingham Malaysia, Jalan Broga, 43500 Semenyih, Selangor, Malaysia

S. V. Mashak
Institute of High Voltage and High Current Faculty of Electrical Engineering, Universiti Teknologi Malaysia, Johor, 81310 Bahru, Malaysia

Z. Zakaria and S. S. Emamian (eds.), *Recent Advances in Electrical and Electronic Engineering and Computer Science*, Lecture Notes in Electrical Engineering 865, https://doi.org/10.1007/978-981-16-9781-4_4

distribution [5, 6]. Electric field nearer to conductor is higher when compared to ground boundary [7].

In this research, FEM is used to simulate the different electric field distribution in high voltage insulators to convert the complicated form over continuous region's problem area into one where it is divided into small elements. FEM is most suitable and commonly used to solve problems involving calculations of electric field and potential distribution in high voltage insulators since this numerical method surpasses others in terms of solving electrostatic problems [8]. This is due to FEM that considers for non-homogeneity of the solution region [9].

2 Methodology

The following equation is used in the simulation of the electric field:

$$E = -\nabla V \tag{1}$$

where E is the electric field distribution, ∇ is the gradient, and V is the calculated voltage. By applying Maxwell's equation,

$$\nabla \cdot E = \nabla \cdot (-\nabla V) = \frac{\rho}{\varepsilon} \tag{2}$$

followed by deriving Poisson's equation to form Laplace equation, the continuity equation to find the contamination of the surface can be found as

$$\nabla^2 \cdot \left(J + \frac{\partial D}{\partial t} \right) = 0 \tag{3}$$

where J is the current density (A/m^2).

Given $J = \sigma E$ and $D = \varepsilon E$ with σ being the electric conductivity (S/m), the equation below is obtained.

$$-\nabla(\sigma \nabla \mathrm{V}) - \nabla \cdot \left(\varepsilon \nabla \frac{\partial D}{\partial t} \right) = 0 \tag{4}$$

For simulation,

$$\nabla^2 V = \frac{\partial^2 D}{\partial x^2} + \frac{\partial^2 V}{\partial y^2} = 0 \tag{5}$$

FEM implements these equations to obtain the area of interest which is the electric field of the insulator.

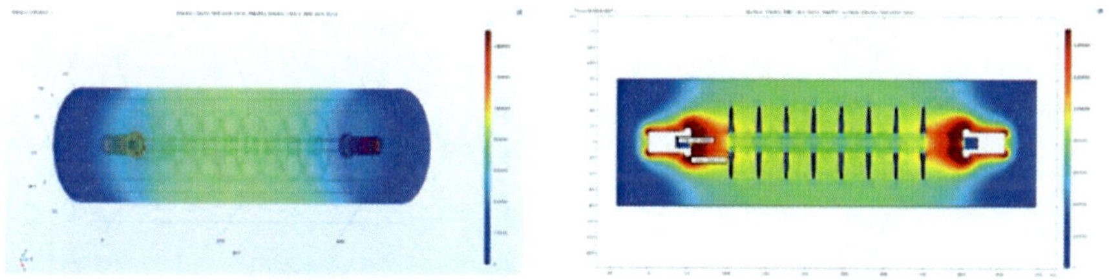

(a) (b)

Fig. 1 Simulation of clean insulator in **a** 3D viewing **b** 2D plane view

With references from [10–12], the geometry of the polymeric insulator is then defined as such that applied voltage = 33 kV, frequency = 60 Hz, r = 0.5 mm, permittivity (uniformly polluted layer) = 81. The 3D modelling was done using AutoCAD and exported into COMSOL. Permittivity of silicone rubber = 4.0, fiber reinforced plastic = 7.1, and air = 1.0 are used for the simulation.

3 Results and Analysis

Two main critical parameters are analyzed using COMSOL, which are the presence of cavity and pollution in polymeric insulator.

3.1 Length

In the clean insulator, the maximum and minimum electric field strength was found to be 485,713 V/m and 10.5888 V/m respectively as shown in Fig. 1.

3.2 Cavity

The electric field strength obtained at the center of the cavity in FRP core and in between FRP core and SRH are 124066 V/m and 114059 V/m respectively as shown in Fig. 2a, b. Figure 2c shows the electric field strength of cavity at SRH which is 36776.4 V/m. Finally, two cavities are simulated, and the results are shown in Fig. 2d. Results obtained show that the more cavities present, the higher the electric field distribution of the insulator inside the cavity.

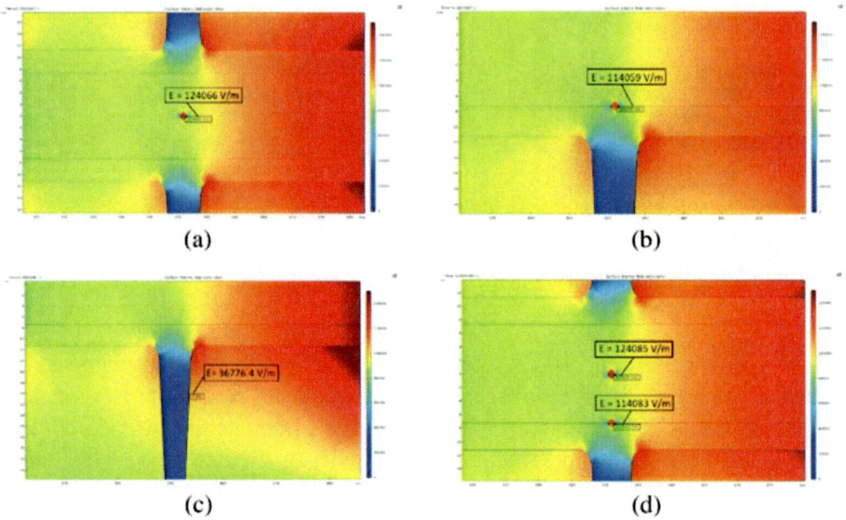

Fig. 2 Simulation of **a** FRP core **b** in between FRP Core and SRH **c** cavity at SRH **d** two cavities

3.3 Pollution

For polluted insulator with cavity in the FRP core as shown in Fig. 3a, there is an increase of 41.92% in electric field strength which is measured to be 188,948 V/m

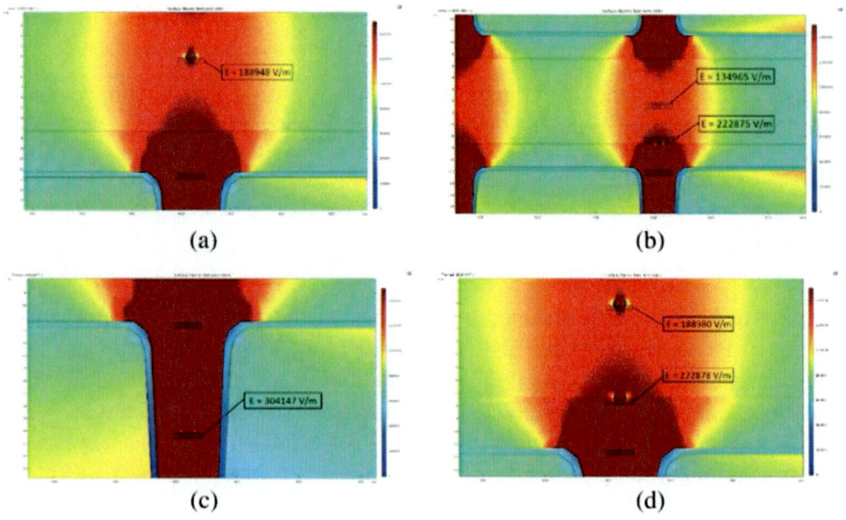

Fig. 3 Simulation of pollution with **a** cavity at FRP Core **b** cavity in between FRP Core and SRH **c** cavity at SRH **d** two cavities

Table 1 Comparison of electric field with uniform pollution layer and cavity at different locations to various other conditions

Comparison	Electric field percentage increase				
	Cavity in FRP with pollution (%)	Cavity in between FRP and SRH with pollution (%)	Cavity in SRH with pollution (%)	Pollution and two cavities (FRP core cavity) (%)	Pollution and two cavities (Cavity in Between) (%)
Pollution + cavity versus Clean Insulator	112.59	167.15	996.90	112.62	167.15
Pollution + cavity versus Pollution	41.92	39.00	33.10	41.94	39.00
Pollution + cavity versus Clean with Cavity	52.30	95.40	727.00	52.32	95.14

compared to only pollution which is only 133,139 V/m at the same location. When compared to clean insulator, an increase of 112.59% is observed.

When the cavity is placed in between the FRP core and insulator shed, there is also a significant increase in the electric field strength observed as shown in Fig. 3b. The electric field strength obtained is 222875 V/m compared to the pollution only which is approximately 67.40% higher. As for the cavity in between the FRP core and SRH, the electric field strength at the center of the cavity is found to be 222,875 V/m.

Pollution with cavity in the insulator shed is simulated and results as shown in Fig. 3c. The electric field strength is found to be 133,138 V/m at the centre of FRP core and 304,147 V/m in the SRH.

Lastly, Fig. 3d shows the simulation of pollution base with two cavities in the polymeric insulator. The results obtained shows that the electric field strength is the highest when two cavities is present in the polluted polymeric insulator which is 188980 V/m or an increase of 41.94%. The centre of the cavity in between FRP and the insulator shed is observed to be 222,878 V/m compared to cavity only which is 114083 V/m. Table 1 summarizes the percentage increases for various conditions from Figs. 2 and 3.

4 Conclusions

The electric field strength of the pollution with presence of two cavities is the highest followed by pollution with cavity in FRP core but is most probably due to the centre of the cavity being the reference point for all the parameters. Therefore, another comparison which was the percentage increase for pollution with cavity in SRH and pollution with cavity in were done. Results shown in both graphs indicate that without cavity, the percentage increase is lower that without pollution. To conclude, the findings in this research indicate that pollution is more severe compared to cavity.

Acknowledgements We would like to thank Swinburne University of Technology, Sarawak Campus for providing the means of resources and for the opportunity to conduct this research project.

References

1. Sadeghi I, Ehya H, Zarandi RN, Faiz J, Akmal AAS (2018) Condition monitoring of large electrical machine under partial discharge fault—a review. SPEEDAM 2018—proceedings international symposis power electron. Electr. Drives, Autom. Motion, 20–22 June 2018, pp 216–223. https://doi.org/10.1109/SPEEDAM.2018.8445261
2. Daphne TYC, Hadi NA, Abdul-Malek Z, Kamyar M, Jubaer A, Tiong SK (2019) Simulation and analysis of electric field distributions in stator bar insulation system with different arrangement of cavities. IEEE, international UNIMAS STEM 12th engineering conference (EnCon), 28–29 August 2019, pp 83–88, 2019. https://doi.org/10.1109/EnCon.2019.8861258
3. Gubanski SM, Dernfalk A, Andersson J, Hillborg H (2007) Diagnostic methods for outdoor polymeric insulators. IEEE Trans Dielectr Electr Insul 14(5):1065–1080. https://doi.org/10.1109/TDEI.2007.4339466
4. Moussavi SZ, Sheikhdoragh B, Shayegani-Akmal AA (2012) Investigation on pollution factors on electric field and potential distribution of polymeric insulator. J Basic Appl Sci Res 2:12482–12491
5. Lopes BRB, Recife B, Ferreira VAL, Recife B, Bezerra JMB (2018) Study of the criticality of internal voids in polymeric insulators with 15 kV voltage class. 2018 Simposio Brasileiro de Sistemas Eletricos (SBSE), pp 1–5. https://doi.org/10.1109/SBSE.2018.8395662
6. Afrouzi H, Abdul-Malek Z, Mashak S, Naderipour A (2013) Three-dimensional potential and electric field distributions in HV cable insulation containing multiple cavities. Adv Mater Res 845:372–377. https://doi.org/10.4028/www.scientific.net/AMR.845.372
7. Afrouzi H, Abdul-Malek Z, Vahabi-Mashak S (2013) Study on effect of size and location of void on electric field and potential distributions in stator bar insulation with finite-element-model. Life Sci J 10:2036–2041
8. Reddy BS, Naik BS, Kumar U, Satish L (2012) Potential and electric field distribution in a ceramic disc insulator string with faulty insulators. 2012 IEEE 10th international conference on the properties and applications of dielectric materials, 24–28 July 2012, pp 1–4. https://doi.org/10.1109/ICPADM.2012.6318928
9. Hani B, M'Ziou N, Boubakeur A (2018) Simulation of the potential and electric field distribution on high voltage insulator using the finite element method. Diagnostyka 19(2):41–52. https://doi.org/10.29354/diag/86414
10. Arshad, Nekahi A, McMeekin SG, Farzaneh M (2015) Effect of dry band location on electric field distribution along a polymeric insulator under contaminated conditions. 2015 50th international universities power engineering conference (UPEC), 1–4 Sept. 2015, pp. 1–4, 2015. https://doi.org/10.1109/UPEC.2015.7339957
11. Muniraj C, Chandrasekar S (2012) Finite element modeling for electric field and voltage distribution along the polluted polymeric insulator. World J. Modell. Simul. 8:310–320
12. Rohit PR, Rahul PR (2018) simulation studies of composite insulators used in high voltage transmission. Int J Eng Res Elect Eng 4:11–17

Modelling and Simulation of Building-Integrated Photovoltaics (BIPV) Installations in Swinburne University

Derisee Tang Shao Ting, Hadi Nabipour Afrouzi, Md. Bazlul Mobin Siddique, Ateeb Hassan, and Jubaer Ahmed

Abstract Governments and organizations around the globe are increasingly searching for ways to reduce pollution from their sectors, such as greenhouse emissions, with a significant focus on the use and implementation of sustainable RE generation. A large portion of renewable energy (RE) work has been devoted to Photo-Voltaic (PV) systems using solar energy to generate electricity to provide both electrical and thermal charges as the abundance, free and clean characteristics of this RE source cause no disturbance or carbon emissions to the environment. Building Integrated Photovoltaic (BIPV) systems have developed tremendously in Malaysia in recent years as an application of PV technology and have been shown to help buildings partially fulfil the load as a feasible RE generation technology. In this paper, the efficiency and feasibility of BIPV application in Swinburne University of Technology Sarawak is investigated. Weather details such as air temperature, relative humidity, daily solar radiation, and the earth temperature in the university will be collected and analyzed. Different BIPV solar power systems are designed and simulated technically through modelling and simulation on related software. The analysis revealed that Feed in BIPV system is consider as the most promising way to achieve the target of saving. The maximum yearly consumption of the building will be 320000 kWh. By considering the BIPV system is connected to the grid, there will be extra 39715 kWh of production being fed back to the grid each year which gives an earning of RM11915 each year. Besides, the investment ratio will be 33.9% and the payback period for the entire BIPV system will be 14 years.

D. T. S. Ting · H. N. Afrouzi (✉) · Md. B. M. Siddique · A. Hassan · J. Ahmed
Faculty of Engineering Computing and Science, Swinburne University of Technology Sarawak, 93350 Kuching, Malaysia
e-mail: hafrouzi@swinburne.edu.my

Z. Zakaria and S. S. Emamian (eds.), *Recent Advances in Electrical and Electronic Engineering and Computer Science*, Lecture Notes in Electrical Engineering 865, https://doi.org/10.1007/978-981-16-9781-4_5

1 Introduction

PV systems used in constructions are characterized into 2 major groups which are, Building-applied Photovoltaic (BAPV) systems and Building-integrated Photovoltaics (BIPV) systems [1–3]. BAPV systems are standard photovoltaic (PV) solar systems that have no significant impact on the features of the building framework, generally installed over the rooftop, while BIPV systems implement the solar cells into the building structure elements [2]. Nowadays, PV integrated buildings use PV cells to replace conventional construction materials in accordance with the building frame, for instances, rooftop, ceilings, balconies, or exteriors [3, 4]. In this paper, the project is going to be implemented on the rooftop of an education building of Swinburne University.

2 Literature Review

2.1 BIPV System Description

BIPV system often known as integrated power storage technologies for building and was divided into 3 categories which are, off-grid systems, on-grid systems and PV-Trombe wall [3]. A complete solar electrical system requires components to generate electricity, convert electricity into alternating electricity that can be used by appliances, store excess electricity safely [5]. A complete BIPV system includes PV modules, inverters, charge regulator and battery bank.

2.2 BIPV Potential in Malaysia

In 2000, first BIPV system has deployed at the general officer's residence in Tenaga Nasional Berhad in Port Dickson, Malaysia with a sizing of 3.15 kWp [1]. The estimated technical potential of BIPV in Malaysia based on accessible building surfaces is approximately 11,000 MWp [1]. Today, there are approximately 400 kWp of on-grid BIPV system are currently installed in Peninsular Malaysia, especially the 362 kWp BIPV at Technology Park Malaysia (TPM) at Bukit Jalil which features 4824 fixed mounted roof modules, an Uninterruptible Power Supply (UPS) battery bank as a generator backup [5, 6]. BIPV has future implementation possibility in Malaysia and is effective in raising consciousness amongst these players in the manufacturing sector concerning environmental sustainability [1, 6].

2.3 Initiatives of BIPV System in Malaysia

Malaysia has also accredited as International Energy Agency-Photovoltaic Power Systems (IEA-PVPS) candidate, while the PVPS participants have carried out a diverse range of collaborative works to apply solar PV to electricity [7, 8]. Malaysian Building Integrated Photovoltaic (MBIPV) projects have been launched and started on July 2005, has finished around December 2010 [7]. MBIPV promotes business growth of PV cells in Malaysia through PV innovations and capability implementation [9].

2.4 Barrier of BIPV in Malaysia

Deployment of BIPV system is considerably low, as BIPV still experiences constraints in Malaysia, though has been in the field for several years. The cost of generating electricity per unit cell is much expensive when compared with conventional fossil or nuclear power generation [9]. Besides, the process of solar cell implementation is complicated as it involved various parameters consideration. At present, BIPV system represents a minor portion of PV industry primarily due to the misconception of BIPV system integrations are costlier than traditional rooftop deployments and that the powerful subsidies delivered from the past are only associated with items of "utility" and non-distributed generation (solar parks) [8, 9]. Typically, the biggest obstacle for emerging new technologies is huge investment costs, which might cause difficulties by introducing BIPV system in the construction sector. [9].

3 Research & Methodology

The research location will be Swinburne University of Technology Sarawak Campus, which is located on the latitude of 1.5322626 and longitude of 110.3550372 in Kuching. BIPV application will be simulated on Building B rooftop. Figure 1a shows the research location which is in Kuching.

3.1 Modelling and Simulation of BIPV Installation

Mathematical modelling such as the solar panel power used, inverter power and the desired daily power demand for the BIPV system will be done by manual calculation. Next, 3D modelling of chosen housing area and building will be built on BIM platform (REVIT), a bridge between design and simulation software for solar energy, and next imported to PYSITES for further simulation. To get a precise weather condition of

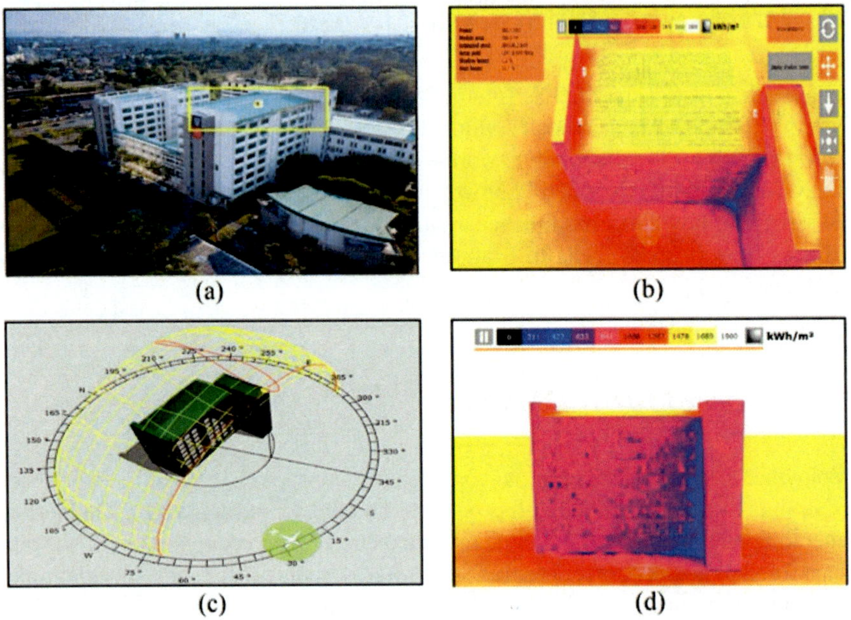

Fig. 1 **a** Research location, **b** Thermal analysis in PVSITES, **c** Sun path view from upward direction analysis in PVSITES, **d** Average yearly irradiation of all hour's analysis in PVSITES

demo site, RETScreen, is used to load the weather details such as thermal gradient, humidity levels, solar daytime radiation and surface temperature in Kuching into a CSV file which will next be loaded into PVSITES. The daily Watt-hours per day to be generated by the building is 884.83 kWh and the calculations below show the targeted daily average power required by Swinburne University of Technology Sarawak Campus:

$$E\,(kWh)\,(assumed\ 10\%\ power\ loss\ in\ transmission) = Daily\ Watt - hours\ per\ day * 1.1 \tag{1}$$

$$Total\ Watt - peak\ rating\ needed\ for\ PV\ modules = \frac{\boldsymbol{Total\ Watt - hours\ per\ day}}{\boldsymbol{Panel\ generation\ factor\ of\ Malaysia}} \tag{2}$$

By using Eq. (1), the targeted average power per day is 973.31 kWh. By substituting, the value into Eq. (2), the total Watt-peak rating needed for PV modules is 286.26 kWp. Next, by considering the panel is not able to generate exact 100% of the power, the total Watt-peak rating will be included an extra 20% of power expansion and given a value of 344 kWp.

4 Results and Discussion

Simulation on PVSITES will be the major perceptions to be analysed in this paper: the 3D model of Swinburne building will be imported and the workflow impact in the solar energy simulation field will be observed in PVSITES.

4.1 BIPV Module Selection

The watt peak per day used in Swinburne building will be 344 kWp. Hence, the solar panel used to be installed on Swinburne building rooftop will need to generate a total power of 40.55 kW daily. The BIPV cell used for this project is Monocrystalline Silicon made BSM500M-96, manufactured by BLUESUN Solar, and with a maximum power of 500 W per cell.

$$Capacity\,of\,inverter = Maximum\ power\ usage + (Maximum\ power\ usage \times 30\%) \quad (3)$$

4.2 Inverter Selection and Wiring

The maximum daily power usage is assumed to be 55 kW since the daily power usage is 40.55 kW and the capacity of the inverter must be 25% to 30% greater than the maximum power usage. Based on Eq. (3), the capacity of inverter is 71.50 kW. The model of the solar inverter is BLUESUN Growatt-18000UE with the maximum DC voltage of 1000 V and minimum 300 VDC. The solar cell specification has summarised in Table 1.

Table 1 Selected solar cell specification from PVSITES

BlueSun—BSM500M-96	
Technology	mono_Si
Peak power	10.0Wp
Size	100 × 100 mm
V_{OCC}	0.61 V
I_{SC}	10.87A
Busbars	5

4.3 BIPV Layout

The total number of the panel can be calculated using the formula (4):

$$\text{Total number of the panel} = \frac{\textbf{Total watt peak per day used(with20\% power expansion)}}{\textbf{Solar panel output}} \quad (4)$$

However, only 228 panels will be installed on the rooftop and generate 30% of the load instead of 715 panels as calculated using the formula due to the insufficient space on the rooftop. Figure 1b shows the thermal analysis of the system in PVSITES. In addition, Fig. 1c shows the Sun path view from upward direction analysis in PVSITES and Fig. 1d shows Average yearly irradiation of all hour's analysis in PVSITES.

4.4 Financial

An operating cost (OPEX) is a compulsory expense for the day-to-day functioning of an enterprise. In contrast, a capital investment (CAPEX) is an investment a business makes to produce a future profit [7]. The estimated value CAPEX and OPEX of the BIPV system will be RM139320 and RM2300.

4.5 Feed in

Feed-in Tariff (FiT) program by following the Net Energy Metering Scheme (NEM) [8]. The key parameters for the system have been summarized and recorded in Table 2.

Table 2 Key parameters for Feed in system

Electricity network	Feed-in tariff	0.3 RM/kWh
	Network tariff	0.2 RM/kWh
Key figures	Total power	285.0 kWp
	Final yield	139.3 kWh/kWp
	Feed-in production	39,715 kWh
	Self-consumed production	0 kWh
	Lost production	0 kWh
	Feed-in income	RM 11,915
	Self-consumption savings	RM 0
	20 years benefit / investment ratio	33.9%
	Payback period	14 years

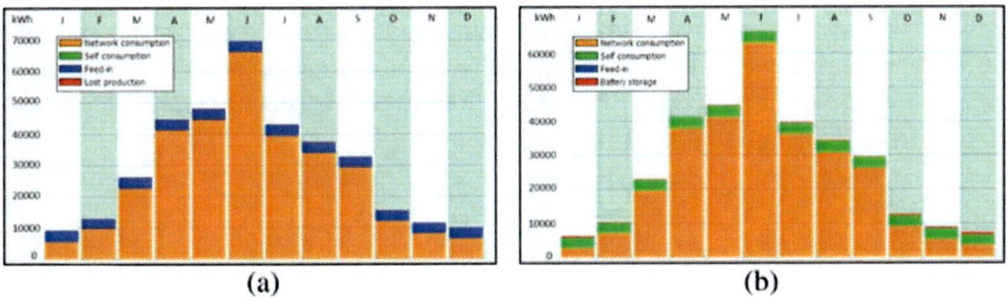

Fig. 2 **a** Energy distribution for Feed in system, **b** Energy distribution for SELCO system

The maximum yearly consumption of the building will be 320000 kWh. There will be extra 39715 kWh of production being fed back to the grid each year and will give an earning of RM11915 each year (the FiT rate for C1 building will be RM0.3/kWh). Besides, the investment ratio will be 33.9% and the payback period for the entire BIPV system will be 14 years. Figure 2a shows the energy distribution of Feed in system and Fig. 2b shows the Energy distribution for SELCO system.

5 Conclusion

Overall, a solar PV system is not feasible to be implemented for all of Swinburne due to the high-power demand and low area to lay all the solar panels. Due to the insufficient space for panel installation, the PV sizing was done for a specific block (Block-B) of Swinburne, which consumes an estimate of 973.31 kWh daily shows that:

- 30% of the power demand is met with 228 of BlueSun Monocrystalline Silicon made panels.
- A yield of 140 kWh/kWp, covering a mounted area of 880 m^2 on the roof tops of Block-B.
- The cost of the Total PV system is around RM200,000.
- The BIPV system has a 14-year payback period and a 33.9% investment ratio over a 20-year period.

Similar BIPV modelling system of 615 m^2 has been simulated on demo site in Ludvika, Sweden, shows that:

- 100% of the power demand is met with 776 of ONYX manufactured BIPV cells.
- A yield of 837 kWh/kWp could yield a maximum of 29,000 kWh.
- The cost of the Total PV system is estimated 35,700€ (RM18000) and maintenance costs of 1785€ (RM9000)
- The BIPV system has a 12-year payback period and a 61% investment ratio over a 20-year period.

It can be concluded that BIPV system simulated on the Ludvika as well as on Swinburne demo building is a feasible project in terms of energy potential and economic investment. Therefore, investing on a Solar PV system is recommended due to the long-term benefits and to help the environment.

References

1. Afrouzia HN, Mashaka SV, Abdul-Maleka Z et al (2013) Solar array and battery sizing for a photovoltaic building in Malaysia. J Teknol (Sci Eng) 64(4): 79–80. https://doi.org/10.11113/jt.v64.2106
2. Nur Huda MB, Afrouzia HN et al (2019) A review of available hybrid renewable energy systems in Malaysia. Int J Power Electron Drive Syst (IJPEDS) 11(1): 433–441. https://doi.org/10.11591/ijpeds.v11.i1.pp433-441
3. Elinwa U, Radmehr M, Ogbeba J (2017) Alternative energy solutions using BIPV in apartment buildings of developing countries: a case study of North Cyprus. Sustainability 9(8):1414. https://doi.org/10.3390/su9081414
4. Nur Huda MB, Afrouzia HN et al (2020) Optimization analysis of hybrid renewable energy system using homer software for rural electrification in Sarawak. International UNIMAS STEM 12th engineering conference (EnCon), IEEE.https://doi.org/10.1109/EnCon.2019.8861260
5. Espeche JM, Noris F, Lennard Z et al (2017) PVSITES: building-integrated photovoltaic technologies and systems for large-scale market deployment. Multidis Digit Pub Inst Proc 1(7):690. https://doi.org/10.3390/proceedings1070690
6. Goh KC, Yap ABK, Goh HH et al (2015) Awareness and initiatives of building integrated photovoltaic (BIPV) implementation in Malaysian housing industry. Procedia Eng 118:1052–1059. https://doi.org/10.1016/j.proeng.2015.08.548
7. Afrouzia HN et al (2021) Sizing and economic analysis of stand-alone hybrid photovoltaic-wind system for rural electrification: a case study Lundu, Sarawak. Clean Eng Technol 4:100191. https://doi.org/10.1016/j.clet.2021.100191
8. Hossain M, Rahim N, Solangi K et al (2011) Global solar energy use and social viability in Malaysia. 2011 IEEE conference on clean energy and technology (CET), 27–29 June 2011, IEEE, pp 187–192. https://doi.org/10.1109/CET.2011.6041461.
9. Palencia PS, Chivelet NM, Chenlo F (2019) Modeling temperature and thermal transmittance of building integrated photovoltaic modules. Sol Energy 184:153–161. https://doi.org/10.1016/j.solener.2019.03.096

Ant Colony Optimization Algorithms for Routing in Wireless Sensor Networks: A Review

R. G. C. Upeksha and W. P. J. Pemarathne

Abstract Wireless Sensor Networks (WSNs) are widely used in applications to extract data from the physical environment and transmit it over different locations using small sensor nodes. The routing protocol of the network plays a major role when transmitting data packets over routing paths. In order to achieve trustworthy wireless communication, it is important to have a reliable, efficient and optimized routing protocol. Ant Colony Optimization Algorithm (ACOA) has been applied by recent researchers to improve and evolve routing algorithms for WSNs, considering ACOA factors like shortest distance, time and number of neighbour nodes. Main goal of this study is to review recently improved Ant Colony Optimization routing algorithms to enhance characteristics in WSNs such as energy-efficiency, Quality of Service (QoS), security and network lifetime.

Keywords Wireless Sensor Networks · Ant Colony Optimization · Routing

1 Introduction

Wireless Sensor Networks are smart network applications which gather, integrate, and communicate data automatically within the network. The WSN is an evolving IT technology integrated with microelectronics, network and communication technologies that delivers the latest technical developments. WSNs are applicable in different areas for various purposes [1]. Application areas can be categorized into major application areas as follows: Agriculture, Environmental Monitoring, Vehicle Tracking, Health Care Monitoring, Smart Building Monitoring, Security Surveillance and Animal Tracking.

A routing technique is essential to identify the neighboring nodes effectively and build a network without the support of an infrastructure to accomplish network connection. In hierarchical networks, several clusters are formed where cluster heads

R. G. C. Upeksha (✉) · W. P. J. Pemarathne
Department of Computer Science, Faculty of Computing, General Sir John Kotelawala Defence University, Rathmalana, Sri Lanka
e-mail: upekshargc@kdu.ac.lk

Z. Zakaria and S. S. Emamian (eds.), *Recent Advances in Electrical and Electronic Engineering and Computer Science*, Lecture Notes in Electrical Engineering 865, https://doi.org/10.1007/978-981-16-9781-4_6

control a group of sensor nodes and routing is simple inside the group. Flat networks store data in neighboring nodes to select paths for routing. However, there are some difficulties found in the WSN routing. For instance, when data is transmitted directly to the sink node, it requires more energy from the sending node. Likewise, for multi-hop networks, the nearest nodes to the sink node face difficulty in forwarding data packets [2].

Similarly, major issues may be present in these important features of energy consumption in nodes and in the whole network, quality of service, security, data transmission and robustness [3]. Algorithms can optimize factors in routing which help to address issues in these functions. For instance, the Ant Colony Optimization (ACO) algorithm has achieved these benefits like finding the shortest path, robustness, distributed computing and so on [4].

In this review article it focuses on application of the ACO algorithm in WSN routing. In addition, how it helps find solutions to issues in WSNs by enhancing the features and how amended or improved ACO algorithms increase the performance of these factors. This study has selected most recent research works for reviewing. The rest of the paper is organized as follows: Sect. 2 briefly introduces the background theories while Sect. 3 describes the characteristics of WSNs. In Sect. 4, review some recently published research findings, Sect. 5 carries the analysis of the review and Sect. 6 presents the conclusion of the paper. The main contributions of this work are as follows:

- Presents some important characteristics of Wireless Sensor Networks that have been widely focused on, when improving ACO routing algorithms.
- Describes some recent improvements in the ACO algorithm that have been used to enhance the performance of WSNs routing and how they work in the routing process.
- Identifies the most focused features in enhancing and the challenges faced during the review study.

2 Background

2.1 Wireless Sensor Networks (WSNs)

Wireless Sensor Networks are mainly consisted of small sensor nodes which are battery powered and communicate within short distances. WSNs can be distributed on a large scale using a big number of sensor nodes. Gathered data is processed in the systems to collect significant information in respective areas [5]. Routing protocols are designed for Wireless Sensor Networks considering factors that are useful to achieve the assurance of efficiency and reliability of the network. Routing techniques and algorithms can be categorized into classical algorithms and Computational Intelligence (CI) based algorithms. The Ant Colony Optimization Algorithm comes under

CI algorithms [6]. ACO has emerged as the most used CI approach by researchers to tackle routing difficulties in WSNs [7].

2.2 *Ant Colony Optimization (ACO)*

Ant colony algorithm is a kind of collective algorithm, which is inspired by the foraging behaviour of ants. Ants are amazingly capable of finding the shortest paths between their nests and the source of food. Actual communication is done by following the pheromone trails laid by an individual ant. Other ants follow the footprint, depending on the intensity of the pheromone trail on the path. If a path has a stronger pheromone trail, there is a high possibility of other ants following this path. This phenomenon automatically leads ants to follow the shortest path.

Marco Dorigo and colleagues introduced the first ACO algorithms in the early 1990s [8]. The observation of ant colonies inspired the development of these algorithms. Ants are social insects and live-in colonies; their behaviour is governed by the goal of colony survival than focused on the survival of individuals. The behaviour that inspired ACO is the ants' foraging behaviour and how ants can find the shortest paths between food sources and their nest.

Ants initially explore the area surrounding their nest randomly when searching for food. Ants can smell pheromone, and while moving, they leave a chemical pheromone trail on the ground. When choosing their way, they tend to choose, in probability, paths marked by strong pheromone concentrations. As soon as an ant finds a food source, it evaluates the quantity and the quality of the food and carries a portion back to the nest. During the return trip, the quantity of pheromone the ant leaves on the ground may depend on the quantity and quality of the food. The pheromone trails will guide other ants to the food source. The indirect communication between the ants via the pheromone trails known as stigmergy enables them to find shortest paths between their nest and food sources [9].

Scientists have studied the behavioural patterns of ants, and now apply the knowledge to solve severe combinatorial optimization problems. Ant colony optimization (ACO) has been used for many computer applications, including routing, assignment, scheduling, subset, machine learning, and bioinformatics problems [9]. In the artificial intelligence perspective, the ACO algorithm is considered as the most successful strands of swarm intelligence [10].

3 Characteristics of WSN

The performance of the wireless sensor networks is measured considering different metrics. According to the research work, it's novelty and improvement may occur in one or many features in the network. This paper reviews recent improvements in the ACO algorithm that are applied to enhance the characteristics of WSNs. The following are the mostly focused features to be enhanced by researchers.

Energy Consumption: Energy consumption can be identified as total network energy consumption and the node energy consumption. An amount of energy is consumed from a sensor node when different functions and events occur [7]. A single node holds a very limited amount of energy. Therefore, it is important to use energy efficient techniques when designing routes in WSNs [11].

Network lifetime: Reliability of the network mainly depends on network lifetime. To maintain the network lifetime at a higher level, it is necessary to keep a balanced energy consumption in the network [7].

Quality of Service (QoS): In WSNs, sensor nodes are dedicated not only for transmitting data through the network but also to monitor the environment to observe events. QoS ensures this transmitting service in the network. Packet loss rate, end-to-end delay and available network bandwidth can be considered as the parameters when determining QoS [12].

Packet Delivery Ratio: The total number of data packets received by the destination is divided by the total number of data packets transmitted from the source yields the packet delivery ratio. To put it another way, the packet delivery ratio is the proportion of packets received at the destination to packets transmitted from the source [13].

Throughput: Average throughput in a network is measured to the ratio of the number of packets received at the destination, to the packet transmission delay [14].

Security: Wireless sensor nodes can be affected by different WSN attacks such as black hole attack, sinkhole attack etc. The trust value of the node is denoted by the reliability of packets or forwarding and receiving packets [15].

4 Recent Ant Colony Optimization Routing Algorithms for WSNs

In this section, authors review some recently proposed routing algorithms that have been improved using the Ant Colony Optimization Algorithm. Rathee and Kumar [16] discussed that most recent research on routing in WSNs have been carried out considering Quality of Service (QoS), Energy Consumption and Security within the network. Therefore, authors have supposed to depict the significance of network lifetime along with these three features. Random occurrences of events in the sensing field, generate routing issues in the network. Authors have proposed an ACO based Quality of Service aware energy balancing secure routing (QEBSR) algorithm to address these censorious problems in WSNs. Trust value in nodes and end-to-end delay has been improved by the proposed algorithm that helps to increase the network lifetime. After the execution of this improved routing algorithm for a preset number of iterations, the best path is given as the result. This final path from source node to sink node updates the residual energy of nodes and the number of dead nodes. This proposed algorithm has been compared with the existing DEBR and EENC algorithms. As a result, network lifetime, Packet Delivery Ratio and reducing delay have all been enhanced when compared to other two algorithms with better performances.

Kaur and Kumar [7] have focused on WSNs in Intelligence Monitoring Systems (IMS). This research work has considered two optimized outcomes in energy consumption and end-to-end delay. Authors have addressed an issue of data traffic in IMS applications that requisite data transmission on time and accurate towards the Base Station (BS). Thus, the energy consumption and end to end delay have been optimized when determining the routing path. Authors have proposed a Multi-objective ACO based QoS-aware Cross-Layer Routing Protocol (MACO-QCR) that makes several contributions to the research problem. In this proposed algorithm, ACO is used to find the optimal routing path considering residual energy in a node. The algorithm selects the node with higher residual energy as the next hop node in the routing path where residual energy in a node act as the pheromone level. Furthermore, higher residual energy emphasizes the higher lifespan of the network. In the analysis, Kaur and Kumar have compared the improved algorithm with existing IAMQER and O_ARA algorithms. It has shown better results in view of reduced latency, increased reliability, and reduced energy consumption in the network.

Arora et al. [17] have illustrated a method for determining the optimum path towards the base station in WSNs. Therefore, a multiple pheromone ant colony optimization (MPACO) technique has been introduced where it is able to find the length between the sensor nodes, residual energy and number of neighbour nodes. Multiple pheromones are used in three ways to figure out the value of these parameters. For instance, from the first pheromone, it can identify the information of neighbour nodes of a one node, such as distance between connected nodes, available nodes, dead nodes etc. The amount of remaining energy or the residual energy can be obtained by the second pheromone. Likewise, the third pheromone is able to find the number of sensor nodes to determine the ideal path. This technique has been compared with recent schemes like LEACH, TBC, PEGASIS and with IACO and MOFCA to contrast the performance. MPACO has given better results in the analysis for enhanced network lifetime and for economical node energy consumption.

Srivastava and Tripathi et al. [14] have focused on designing an energy efficient and network life spanned wireless sensor network by introducing a novel congestion control method. In this proposed mechanism, hybrid K-means and Greedy Breadth First Search algorithms are used to form clusters in the network while the firefly optimization algorithm reduces packet rate. Ant Colony Optimization routing is used to find the optimal path and to increase the throughput. Authors have compared this proposed protocol (AWFCC) with existing research works such as PPI, CDTMRLB and FBCC. Throughput analysis for this routing protocol has shown better performance when compared to the above strategies.

Guleria and Verma [13] have proposed an approach to decide the optimal path in Wireless Sensor Networks using Meta-Heuristic Ant Colony Optimization (MHACO). Furthermore, it requires reliable and accurate event detection to maintain network connectivity. To provide this, it consumes some amount of node energy, and it is important to keep a higher level of residual energy for optimum routing at the same time. In this approach, authors have introduced an intermediate node, namely Rendezvous node (Rnode) that helps to lower the energy consumption in nodes by reducing data transmissions. Hence, the residual energy is increased and helps the algorithm to discover the best routing path. Authors have compared this novel approach with existing methods such as EAUCF, CHEF, FAMACROW and IFUC. The proposed algorithm has shown better performance in packet delivery ratio, reduced energy consumption, residual energy, number of dead nodes and number of received packets by BS.

Arora et al. [18] have focused on the limitations that can be encountered when using traditional routing protocols. Thus, it is difficult to expand the network and dynamic conditions due to high routing costs. Authors have proposed an ACO optimized self-organized tree-based energy balance algorithm for wireless sensor network (AOSTEB) that finds the optimal route for inter-cluster and intra-cluster communication. AOSTEB works in three phases known as cluster formation, multi-path construction and data transmission. In the cluster formation phase, it defines the Cluster Heads (CH) within the network. Thereafter, this proposed protocol finds the path that has the smallest number of intermediate nodes by executing the ACO algorithm on every node. It finds the best neighbour node in the multi-path construction phase. The use of the ACO algorithm helps to determine the most efficient optimized path according to the time that pheromone lasts in a link. Authors have compared this approach with existing IACO, LEACH, PEGASIS, TBC, GSTEB, HEEMP and MOFCA routing algorithms. As the result of this research work, authors have been able to achieve the aims of enhanced network lifetime, reduced energy consumption and scalability with better performances.

Sun et al. [15] have presented a secure routing protocol based on multi-objective ant-colony-optimization for wireless sensor networks (SRPMA) considering the network security along with reduced energy consumption. According to the trust value of the node, the algorithm is able to discover the most secured routing path, avoiding malicious nodes affected by different types of WSN attacks. Improvements in network security consume some amount of node energy where it can lead to high energy consumption. Therefore, authors have proposed this multi-objective optimization in security along with reducing energy consumption of the sensor nodes. Authors have contrasted the performance of SRPMA with a similar existing algorithm, namely IASR. The packet loss rate has been reduced for the SRPMA when the network is affected by a black hole attack, when compared to the IASR.

Sharmin et al. [19] have addressed the challenge facing WSNs when discovering the optimal routing path with less energy consumption. In this research work, authors have extended the previous research work focusing on network performance, which has introduced a novel routing approach based on ACO. The optimal path is selected by this algorithm considering important communication factors such as residual energy, constrained energy, distance between source node and base station when transmitting data. According to the analysis and simulation results, the proposed algorithm has shown least energy consumption and enhanced network lifetime. This algorithm has been compared with the traditional ACO algorithm and EICAntS algorithm for the above performance metrics.

Grosso et al. [20] have focused mainly on many-to-many communication mechanisms for WSNs, which are not widely discussed in research studies. For instance, there are multiple source nodes that can collect data from the environment at the same time and transmit it to many sinks. Issues that can be encountered in the above scenario are network scalability and network lifetime. Authors have addressed these problems by proposing an improved routing algorithm based on ACO. Ants have been considered as packets that transmit over the network. In the many-to-many communication method, many ants are forwarded by sources within a certain period. Authors have developed the traditional ACO algorithm to combine ants into one ant when they meet on the same node. After that, they share a common path named as "backbone" until they split again to find their sink nodes. By introducing this method, the count of packets sent is reduced, that effects enhancing the network lifetime. At the same time, network scalability has increased with shortened backbone length. Compared to a similar network that is tested with flooding protocol, authors have been able to show better performances.

Kumar et al. [21] have presented an algorithm that can detect the path of a mobile sink to collect data from sensor nodes. The mobile sink facilitates traverse only through the Rendezvous Points (RPs). The authors' goal was to improve the ACO algorithm to discover the optimal path from the selected set of optimal RPs. Moreover, the improved algorithm (ACO-MSPD) presents a mechanism to re-select RPs in a dynamic manner to balance the sensor node's energy consumption. Authors have compared the proposed work with some existing algorithms EAPC, WRP, CB and brute force algorithm. The results have shown better performance in network lifetime, energy consumption and packet delivery ratio.

Sun et al. [22] have addressed the security concern in WSNs by applying the D-S evidence theory in routing. WSNs can consist of malicious nodes within the network that have severe effects on data transmitting. It can reduce received and forwarding packets and packet reliability. A trust evaluation model has been established adapting D-S evidence theory. It assesses the node considering both direct and indirect trust values. After that, the ant colony algorithm and trust evaluation model have been integrated to build a secured and trustworthy route in WSNs, avoiding threats from internal malicious attacks. In this combination, trust values are used as the heuristic function of the ACO algorithm. It selects the highly trusted nodes to discover the routing path. The proposed algorithm has been compared with EEABR and IEEABR algorithms to contrast performance determining values for metrics like packet loss rate, end-to-end delay etc.

Zou and Qian [23] have been proposed an Improved Ant Colony Optimization (IACO) algorithm that has focused on solving an issue found in traditional ACO algorithm. In ACO, it tends to define the shortest path from the set of neighbour shortest paths rather than discovering it in the global level. Hence, it takes more time to link up the shortest paths in WSNs. The presented algorithm considers the whole network when transmitting data and communicating with hop nodes. Transferring data in the routing maintenance stage, enhances the pheromone level. The concentration of pheromone helps to find the shortest path. In this research, both shortest path and residual energy have considered when discovering the optimal routing path.

IACO algorithm has been compared with previously developed ACO, DEBA and Dijkstra algorithms under three QoS levels. end-to-end delay, packet loss rate and node residual energy have been compared for these algorithms.

5 Result and Analysis

Table 1 in the review shows a summarization of recent research works. By observing the details, this study can emphasize some features that most authors have considered when they are improving the ACO algorithm. Many authors have focused on reducing energy consumption in the network, increasing network lifetime and security. Therefore, it is able to figure out the importance of reduced energy consumption, extended network lifetime and security of the network when designing a routing algorithm. According to the analyses of these research works, it is highly depicted the contribution of the ACO algorithm to enhance these characteristics in WSNs. Simultaneously, the enhancement of these features leads to an efficient and reliable Wireless Sensor Network. Most authors have considered the pheromone level or updated pheromone value to determine the residual energy of a node or the total path energy, while few authors have used it for obtaining more parameters. Moreover, discovering an optimal routing path was the main goal in most research works reviewed here. The challenge faced when reviewing is the classification of WSN characteristics and performance metrics. Some authors have considered some characteristics as performance metrics and vice versa. However, focusing on both has provided good results in the research outputs. Ultimately, it emphasizes that new improvements and alterations in the ACO algorithm provide very effective and efficient results in routing for Wireless Sensor networks.

6 Conclusion

Wireless Sensor Networks have become a very useful and effective method that is applicable in different physical environments to collect data, transmit data over locations and process. Routing in WSNs is one of the main tasks that should be done in a timely and reliable manner. For instance, some applications like fire detection, gas leakage detection, medical emergencies need rapid and accurate message transmission to take next actions. Hence, data transmitting routes should be more efficient and reliable. The Ant Colony Optimization Algorithm is a very successful study that comes under Swarm Intelligence. It facilitates finding the optimum path between two locations using behavioral patterns of ants. This review presents the recent research works where the traditional ACO algorithm has been improved and applied in routing of WSNs. The study has been able to review and summarize the enhanced features of WSNs using ACO-based improved routing algorithms such that reduced energy consumption, network lifetime, QoS, security, packet delivery ratio

Table 1 Analysis of proposed routing algorithms in the recent literatures

Authors	Proposed algorithm name	Optimization objectives based on ACOA	Determined parameters by pheromone	Enhanced characteristics of WSNs	Routing structure
Rathee et al. [16]	QEBSR	End to end delay, trust value	Residual energy, number of dead nodes	QoS, security	–
Kaur and Kumar [7]	MACO-QCR	Network energy consumption, end to end delay	Residual energy of a node	Network life time, QoS, reduced event traffic	Cross- layer
Arora et al. [17]	MPACO	Node energy consumption, optimum path	Distance between nodes, number of neighbor nodes, residual energy	Economical energy consumption, network lifetime	Self- organizing tree
Srivastava and Tripathi [14]	AWFCC	Shortest path	Path energy	Throughput	Clustering
Guleria and Verma [13]	MHACO	Optimal path, novel cluster head (Ch) selection	Residual energy	Reduced energy consumption, packet delivery ratio, QoS	Unequal clustering
Arora et al. [18]	AOSTEB	Multiple path selection between CH and member nodes, shortest distance	Path Energy	Reduced energy consumption, network lifetime, scalability	Self- organizing tree
Sun et al. [15]	SRPMA	Residual energy, trust value of the path	Residual energy trust value	Security, reduced energy consumption	–
Sharmin et al. [19]	Proposed algorithm	Residual energy, shortest distance	Path energy	Reduced Energy consumption, network lifetime, scalability	Clustering

(continued)

Table 1 (continued)

Authors	Proposed algorithm name	Optimization objectives based on ACOA	Determined parameters by pheromone	Enhanced characteristics of WSNs	Routing structure
Grosso et al. [20]	Proposed algorithm	Combine ants to one, use common path (Backbone)	Total forward route length	Scalability, network lifetime	Many-To-many
Kumar et al. [21]	ACO-MSPD	Optimal RP path	Distance between nodes	Network lifetime, reduced Energy Consumption, Packet Delivery Ratio	Spanning tree
Sun et al. [22]	Proposed algorithm	Secured path	Trust values	Security, Reduced Energy Consumption	–
Zou and Qian [23]	IACO	Shortest path	Distance between nodes	QoS, reduced energy consumption, network lifetime	–

and so on. Finally, it can conclude with clear evidence that the use of Ant Colony Algorithm in WSN routing has made a higher-level contribution to the efficacy of WSN applications.

References

1. Prathap U, Shenoy PD, Venugopal KR, Patnaik LM (2012) Wireless sensor networks applications and routing protocols: survey and research challenges. In: 2012 international symposium on cloud and services computing, Dec 2012, pp 49–56. https://doi.org/10.1109/ISCOS.2012.21.
2. Correal N, Patwari N (2001) Wireless sensor networks: challenges and opportunities
3. Sharma S, Bansal RK, Bansal S (2013) Issues and challenges in wireless sensor networks. In: 2013 international conference on machine intelligence and research advancement, Dec 2013, pp 58–62. https://doi.org/10.1109/ICMIRA.2013.18
4. Luo L, Li L (2012) An ant colony system based routing algorithm for wireless sensor network. In: 2012 international conference on computer science and electronics engineering, Mar 2012, vol 2, pp 376–379. https://doi.org/10.1109/ICCSEE.2012.145
5. Pavai K, Sivagami A, Sridharan D (2009) Study of routing protocols in wireless sensor networks. In: 2009 international conference on advances in computing, control, and telecommunication technologies, Trivandrum, Kerala, Dec 2009, pp 522–525. https://doi.org/10.1109/ACT.2009.133
6. Daanoune I, Baghdad A, Ballouk A (2019) A comparative study between ACO-based protocols and PSO-based protocols in WSN. In: 2019 7th mediterranean congress of telecommunications (CMT), Fès, Morocco, Oct 2019, pp 1–4. https://doi.org/10.1109/CMT.2019.8931320

7. Kaur T, Kumar D (2021) MACO-QCR: multi-objective ACO-based QoS-aware cross-layer routing protocols in WSN. IEEE Sens J 21(5):6775–6783. https://doi.org/10.1109/JSEN.2020.3038241
8. Dorigo M, Maniezzo V, Colorni A (1996) Ant system: optimization by a colony of cooperating agents. IEEE Trans Syst Man Cybern Part B Cybern 26(1):29–41. https://doi.org/10.1109/3477.484436
9. Pemarathne WPJ, Fernando TGI (2016) Wire and cable routings and harness designing systems with AI, a review. In: 2016 IEEE international conference on information and automation for sustainability (ICIAfS), Dec 2016, pp 1–6. https://doi.org/10.1109/ICIAFS.2016.7946575
10. Ant Colony Optimization | MIT Press eBooks|IEEE Xplore. https://ieeexplore.ieee.org/book/6267250. Accessed 09 Jul 2021
11. Zeb A et al (2016) Clustering analysis in wireless sensor networks: the ambit of performance metrics and schemes taxonomy. Int J Distrib Sens Netw 12(7):4979142. https://doi.org/10.1177/155014774979142
12. Lin F, Zhang D, Li W (2011) Research on quality of service in wireless sensor networks. In: 2011 IEEE 2nd International Conference on Software Engineering and Service Science, Jul 2011, pp 312–315. https://doi.org/10.1109/ICSESS.2011.5982316
13. Guleria K, Verma AK (2019) Meta-heuristic ant colony optimization based unequal clustering for wireless sensor network. Wirel Pers Commun 105(3):891–911. https://doi.org/10.1007/s11277-019-06127-1
14. Srivastava V, Tripathi S, Singh K, Son LH (2020) Energy efficient optimized rate based congestion control routing in wireless sensor network. J Ambient Intell Humaniz Comput 11(3):1325–1338. https://doi.org/10.1007/s12652-019-01449-1
15. Sun Z, Wei M, Zhang Z, Qu G (2019) Secure routing protocol based on multi-objective ant-colony-optimization for wireless sensor networks. Appl Soft Comput 77:366–375. https://doi.org/10.1016/j.asoc.2019.01.034
16. Rathee M, Kumar S, Gandomi AH, Dilip K, Balusamy B, Patan R (2021) Ant colony optimization based quality of service aware energy balancing secure routing algorithm for wireless sensor networks. IEEE Trans Eng Manag 68(1):170–182. https://doi.org/10.1109/TEM.2019.2953889
17. Arora VK, Sharma V, Sachdeva M (2020) A multiple pheromone ant colony optimization scheme for energy-efficient wireless sensor networks. Soft Comput 24(1):543–553. https://doi.org/10.1007/s00500-019-03933-4
18. Arora VK, Sharma V, Sachdeva M (2019) ACO optimized self-organized tree-based energy balance algorithm for wireless sensor network: AOSTEB. J Ambient Intell Humaniz Comput 10(12):4963–4975. https://doi.org/10.1007/s12652-019-01186-5
19. Sharmin A, Anwar F, Motakabber SMA (2019) Energy-efficient scalable routing protocol based on ACO for WSNs. In: 2019 7th International Conference on Mechatronics Engineering (ICOM), Putrajaya, Malaysia, Oct 2019, pp 1–6. https://doi.org/10.1109/ICOM47790.2019.8952053
20. Grosso J, Jhumka A, Bradbury M (2019) Reliable many-to-many routing in wireless sensor networks using ant colony optimization. In: 2019 15th European Dependable Computing Conference (EDCC), Naples, Italy, Sep 2019, pp 111–118. https://doi.org/10.1109/EDCC.2019.00030
21. Kumar PD, Amgoth T, Annavarapu CSR (2018) ACO-based mobile sink path determination for wireless sensor networks under non-uniform data constraints. Appl Soft Comput 69:528–540, Aug 2018. https://doi.org/10.1016/j.asoc.2018.05.008
22. Sun Z, Zhang Z, Xiao C, Qu G (2018) D-S evidence theory based trust ant colony routing in WSN. China Commun. 15(3):27–41. https://doi.org/10.1109/CC.2018.8331989
23. Zou Z, Qian Y (2019) Wireless sensor network routing method based on improved ant colony algorithm. J Ambient Intell Humaniz Comput 10(3):991–998. https://doi.org/10.1007/s12652-018-0751-1

Analyzing the Effects of Corona Ring Material and Dimensions on the Electric Field Distribution of 132 kV Glass Insulator String Using 2-D FEM

O. D. Xavier, N. A. Hadi, H. Ateeb, A. Jubaer, M. Kamyar, and A. M. Zulkurnain

Abstract Insulators are highly important in high voltage transmission lines for their ability to regulate the amount of current flowing through the entire line. However, these insulators are vulnerable to electrical discharges such as current leakage, and corona under high voltage conditions. These phenomena cause the insulators to deteriorate overtime which makes them susceptible to breaking down. The paper analyses the effects of corona ring material and dimensions on the electric field distribution of the 132 kV glass insulators. The electric field distribution was calculated using a numerical method known as Finite Element Method where the analysis was done with its corresponding software, and the simulation utilizes a two-dimensional insulator string model. The results show that the installation of corona rings significantly improves the electric field distribution of the insulator string leading to a decrease in electric field of the highly stressed regions by 75%. Meanwhile, a lower installation height for the corona ring would improve the overall electric field distribution of the insulators. A higher installation height would cause the electric field to be less evenly

O. D. Xavier · N. A. Hadi (✉) · H. Ateeb · A. Jubaer
Faculty of Engineering, Computing and Science, Swinburne University of Technology Sarawak, 93350 Kuching, Malaysia
e-mail: hafrouzi@swinburne.edu.my

O. D. Xavier
e-mail: 100069221@students.swinburne.edu.my

H. Ateeb
e-mail: ahassan@swinburne.edu.my

A. Jubaer
e-mail: jahmed@swinburne.edu.my

M. Kamyar
Faculty of Science and Engineering, Department of Electrical and Electronic Engineering, University of Nottingham Malaysia, Jalan Broga, 43500 Semenyih, Selangor, Malaysia
e-mail: kamyar.mehranzamir@nottingham.edu.my

A. M. Zulkurnain
Institute of High Voltage and High Current Faculty of Electrical Engineering, Universiti Teknologi Malaysia, 81310 Bahru, Johor, Malaysia
e-mail: 100069221@students.swinburne.edu.my

Z. Zakaria and S. S. Emamian (eds.), *Recent Advances in Electrical and Electronic Engineering and Computer Science*, Lecture Notes in Electrical Engineering 865, https://doi.org/10.1007/978-981-16-9781-4_7

distributed along the string. On the other hand, a higher ring diameter would worsen the field distribution where there was an increase in electric field of up to 54.1% at the bottom and the top of the string.

1 Introduction

Insulators play an important part in transmission lines as their insulation performance dictates the reliability of a transmission network [1, 2]. Not only do they provide mechanical support on the suspension of overhead transmission lines, they also isolate live phase conductors from the support towers [3]. Insulator strings are usually designed to minimize the build-up of electrical charges around its surfaces which aims to reduce the frequency of current leakage, arcs and flashovers. A high voltage transmission line is prone to electrical discharges that may eventually lead to the breakdown of the transmission network if not treated properly [4].

The state of the electric field distribution on a high voltage component usually dictates the frequency of electrical discharges around it and the performance of the insulators thus making it highly important in designing equipment in high voltage levels [5]. A non-uniform electric field will cause more flashovers to occur that may degrade the insulators [6, 7]. To reduce the frequency of the phenomena, corona rings are usually installed on the insulator strings as it improves the insulation performance of the string [8]. To maximize the effect of the corona rings, the electric field distribution around the high voltage component needs to be studied. There are two methods that have been devised to study this problem where the first is to conduct an experiment in a controlled environment with proper equipment. This method is generally time consuming and tedious to set up which is the second method is preferable. The second method is the utilization of computational programs to simulate the electric field distribution around the insulator strings along with the effects of corona ring. There are many engineering computer programs that can be used for simulation if it utilizes the method of interest. For this case, finite element method (FEM) was used. Finite element method is a numerical method introduced in 1960 and initially applied in electrical engineering in 1965 [3]. FEM can be used to describe the behaviour of electromagnetic fields through modelling. While there are other numerical methods that can be used to calculate the electric field distribution such as charge simulation method (CSM), boundary element method (BEM) and finite difference method (FDM), FEM is chosen as that is the method of choice for this topic [9].

2 Electric Field and Potential Formulation

The basic formulas used to calculate the electric field distribution involves the use of Maxwell's equation [10]. The partial differential equation that defines the voltage potential distribution is defined below:

$$\nabla.\boldsymbol{D} = \rho_v \tag{1}$$

where $\boldsymbol{D}$ is the electric flux density (Cm^{-2}) and ρ_v is the free charge density (Cm^{-3}). Given that $\boldsymbol{D} = \varepsilon \boldsymbol{E}$ where $\varepsilon = \varepsilon_0 \varepsilon_r$, Eq. (1) becomes:

$$\nabla \cdot (\varepsilon_0 \varepsilon_r \mathbf{E}) = \rho_v \tag{2}$$

where E is the electric field, ε_0 is permittivity in vacuum and ε_r is the relative permittivity. The electric field E can be defined in terms of voltage potential, V as the:

$$\boldsymbol{E} = -\nabla \mathrm{V} \tag{3}$$

Substituting (3) into (2) will create the following equation:

$$\nabla \cdot (\varepsilon_0 \varepsilon_r (-\nabla \mathrm{V})) = \rho_v \tag{4}$$

After expanding the equation and rewriting it, Poisson's equation was obtained.

$$\nabla \cdot \nabla \mathrm{V} = -\frac{\rho_v}{\varepsilon_0 \varepsilon_r} \tag{5}$$

In the absence of free charges where $\rho_v = 0$, Laplace's equation is obtained. Laplace's equation expressed in Cartesian coordinates would be as shown below:

$$\frac{\delta^2 V}{\delta^2 x} + \frac{\delta^2 V}{\delta^2 y} + \frac{\delta^2 V}{\delta^2 z} = 0 \tag{6}$$

Equation (6) only applies to homogenous mediums within a region. For non-homogenous medium, the equation can be rewritten as:

$$\varepsilon_0 \varepsilon_r \left(\frac{\delta^2 V}{\delta^2 x} + \frac{\delta^2 V}{\delta^2 y} + \frac{\delta^2 V}{\delta^2 z} \right) = 0 \tag{7}$$

Further derivations and calculations were handled by the FEM-based software.

3 Parameters of Insulators and Corona Ring Profiles

A 10-unit glass insulator string was used as the model, and 10 different profiles of corona ring was studied. Six case studies were done to observe the effects of corona ring material and dimensions where the corona ring with profile A served as a reference model for comparison. The parameters of the glass insulator and corona

Table 1 Design parameter for the suspension-disc glass insulator

Dimensions (mm)	
Diameter (D)	175
Spacing (H)	110
Creepage distance	200
Coupling size	11
Material	Glass

Table 2 Relative permittivity for all materials used in the simulation

Material	Relative permittivity
Stainless steel	1
Vacuum	1
Glass	5
Alumina (Al_2O_3)	9.5
Free space	8.85×10^{-12} F/m

rings are shown in Tables 1, and 3, while the relative permittivity for all materials used is shown in Table 2. The presence of a pollution layer and the model for a transmission tower was not considered. The simulation was conducted as a static study where variables such as time and frequency were not considered.

4 Results and Discussion

4.1 Corona Ring Effect

The electric potential shown in Fig. 1 indicates the effect of installing corona rings on the insulator string. Without corona rings, the potential is highly concentrated at the bottom part of the string. With the corona rings, the potential is more evenly distributed throughout the insulator string.

Table 3 Corona ring parameters for all 10 profiles for their respective case study

Parameters (mm)	Profiles									
	A	B	C	D	E	F	G	H	I	J
Installation height	100	25	100	100	100	300	100	100	100	100
Ring diameter	300	300	600	300	300	300	800	300	300	300
Ring tube diameter	50	50	50	60	50	50	50	40	50	50
Inner tube diameter	25	25	25	25	10	25	25	25	40	25
Material	Stainless steel									Alumina

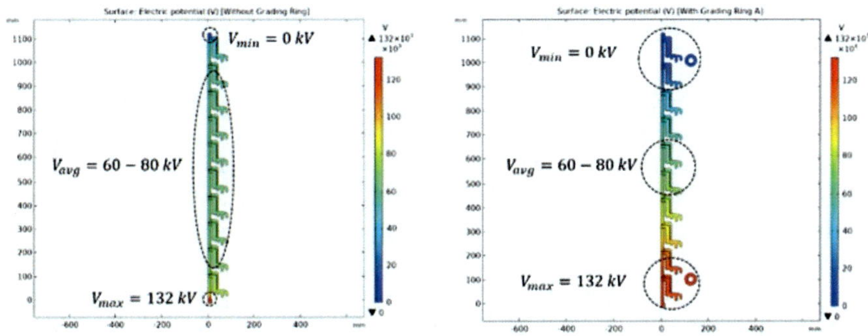

Fig. 1 Electric potential of the glass insulator string without corona ring (left) and with corona ring A (right)

Figure 2 shows the electric field distribution for both cases where the field distribution at the bottom and top of the string with corona ring A is 75% lower compared to the string without the rings. Meanwhile, the field distribution for the middle region of the string with corona ring A shows an increase, ranging of 42.5–784% compared that the string without the ring.

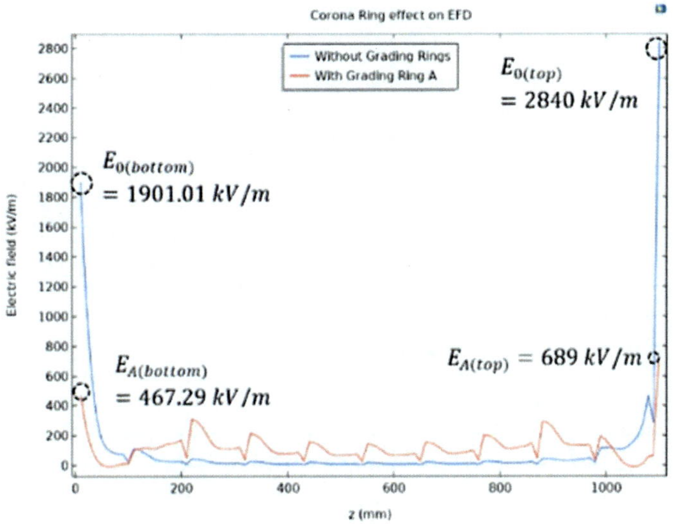

Fig. 2 Electric field distribution of the glass insulator string without corona ring and with corona ring A

4.2 Installation Height Effect

With corona ring A as reference for comparison, Fig. 3 shows the effect of varying the ring's installation height on the insulator string. Comparing all three profiles, a smaller installation height for the string with corona ring B decreases the electric potential around the lowermost insulator units while a bigger installation height for the string with corona ring F shifts the concentration of electric potential to the middle insulator units of the string.

The electric field distribution for the insulator string with ring B is lower compared to the string with Ring A by 9–28.7% as shown in Fig. 4. The only increase in

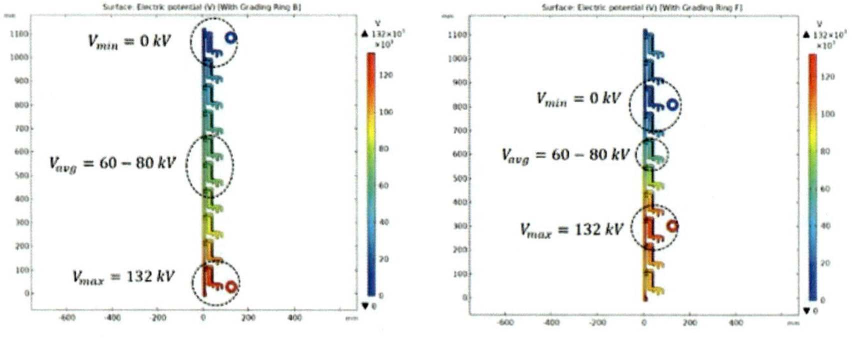

Fig. 3 Electric potential of the glass insulator string with corona ring B (left) and with corona ring F (right)

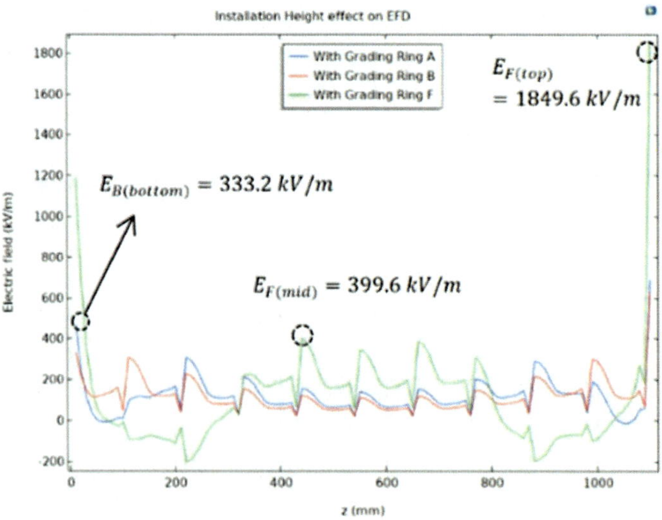

Fig. 4 Electric field distribution of the glass insulator string with corona ring A, B and F

field distribution is located at the 100 and 1000 mm mark where it is 380.4–73.8% higher. On the other hand, the field distribution for the string with grading ring F is significantly higher at the bottom and top part with an increase of 154.1% and 168.2% respectively. The region between 400 and 800 mm has a higher electric field distribution by approximately 150% while the field distribution at the 200 and 900 mm-mark is lower by 167%.

4.3 Ring Diameter and Ring Tube Diameter Effect

The insulators in Fig. 5 possess higher ring diameters than the insulator string with ring A. It is visible that the increase in ring diameter decreases the electric potential on the surface of the string. The increase in ring diameter equates to the increase in distance between the corona rings and the insulator string.

The left graph in Fig. 6 shows that the electric field distribution for the model with ring C is lower on other regions except the bottom and the top of the string where the field distribution is 27.5 and 37.2% higher than the model with ring A. Meanwhile, the center region exhibits a slight increase of electric field by 28% while the electric field around the 200 and the 1000 mm region is lower than the model with ring A by 40%. The electric field for the string with ring G shows similar traits to the model with ring C with the only notable difference being the field at the bottom and the top region being higher than that of ring A by 43% and 54.1% respectively. On the other hand, the right graph in Fig. 6 shows how minor the effect of varying ring tube diameters can be on the electric field distribution of the insulator string. The electric field at the lowermost and uppermost unit of the string with ring H is higher than ring A by 16.9–19.2%. Meanwhile, the electric field at the lowermost and uppermost unit for the string with ring D is lower than ring A by 15.1–17.7%. Otherwise, the change in electric field for the other parts of the string is relatively small with some regions

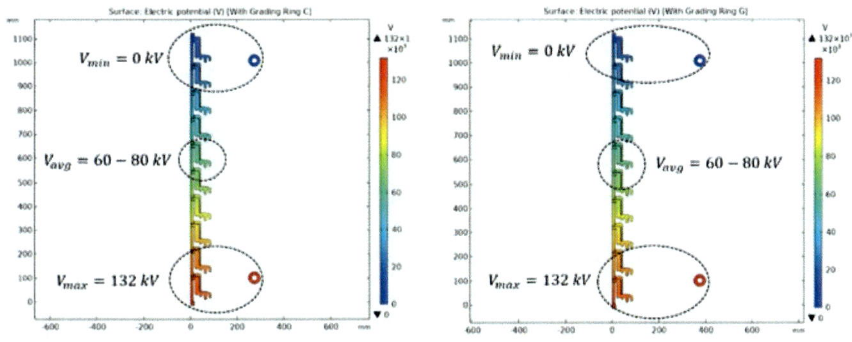

Fig. 5 Electric field distribution of the glass insulator string with corona ring C (left) and with corona ring G (right)

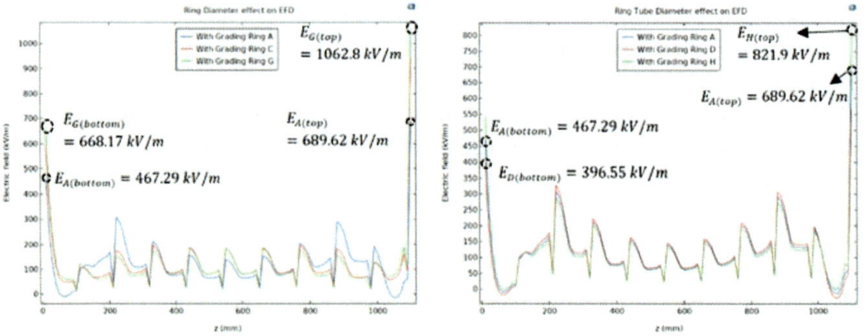

Fig. 6 Electric field distribution of the glass insulator string showing the effect of varying ring diameter (left) and varying ring tube diameter (right)

seeing an increase, averaging at 4.51% for the model with ring D and a decrease in electric field for the model with ring H, averaging at 4.83%.

4.4 Inner Tube Diameter and Ring Material Effect

As shown in Fig. 7, there is no noticeable change in the electric potential along the insulator string regardless of the change in inner tube diameter. The same can be said to the change in corona ring material. It is safe to conclude that varying inner tube diameter does not provide any benefits for the electric field distribution of the insulator string. It is worth noting that the value of the inner tube diameter is limited to the outer ring tube diameter and it is not logical for the inner tube diameter to exceed that of the outer ring tube diameter. The difference in corona ring material has no visible effect on the electric potential and the field distribution of the insulator

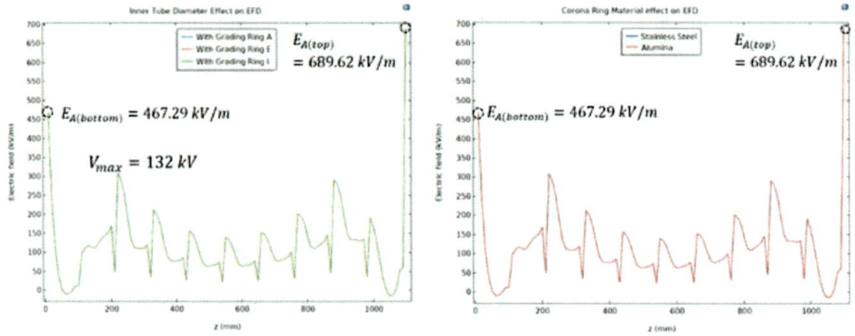

Fig. 7 Electric field distribution of the glass insulator string showing the effect of varying inner tube diameter (left) and varying corona ring material (right)

string and the difference in relative permittivity does not bring a noticeable change in the electric potential.

5 Conclusion

To summarize, the modelling of a two-dimensional model of a 10-unit glass insulator string along with the effect of corona rings, its design parameter and material was studied. The installation of corona rings significantly improves the electric field distribution of the insulator string leading to a decrease in electric field of highly stressed regions by 75%. A lower installation height for the corona ring would improve the overall electric field distribution of the insulators and the increase in ring diameter would worsen the field distribution where there was an increase in electric field of up to 54.1% at the bottom and the top of the string. For the ring tube diameter, a higher tube diameter would provide a better field distribution while a lower tube diameter would worsen the field distribution. Finally, the ring's material and inner tube diameter does not show a significant change in the field distribution of the string. It is possible to obtain results with higher accuracy by using a three-dimensional model in the simulation and introducing further constraints to the simulation such as the conductivity of the material.

Acknowledgements Authors wish to thank Swinburne University of Technology for the ability to complete the research project.

References

1. Othman NA, Piah MAM, Adzis Z et al (2014) Characterization of charge distribution on the high voltage glass insulator string. J Electrostat 72(4):315–321. https://doi.org/10.1016/j.elstat.2014.05.003
2. Afrouzi HN, Abdul-Malek Z, Mashak SV (2013) Study on effect of size and location of void on electric field and potential distributions in stator bar insulation with finite-element-model. Life Sci J 10(4):2036–2041
3. Akbari E, Mirzaie M (2012) Investigating the effects of disc insulator type and corona ring on voltage distribution over 230-kv insulator string using 3-d FEM. Int J Eng Sci Emerg Technol 3(1):1–8
4. Ilhan S, Ozdemir A (2007) Effects of Corona ring design on electric field intensity and potential distribution along an insulator string. In: 7th international conference on electrical and electronics engineering, December 5–9, vol 1, pp 142–146
5. Afrouzi HN, Abdul-Malek Z, Vahabi Mashak S et al (2014) Three-dimensional potential and electric field distributions in hv cable insulation containing multiple cavities. Adv Mater Res 845(2014):372–377. https://doi.org/10.4028/www.scientific.net/AMR.845.372
6. Krzma AS, Khamaira MY, Abdulsamad M (2018) Comparative analysis of electric field and potential distributions over porcelain and glass insulators using finite element method. In: Proceedings of first conference for engineering sciences and technology, September 25–27, vol 1, pp 177–183

7. Chee DTY, Afrouzi HN, Abdul Malek Z et al (2019) Study of electric field distribution in the high voltage stator bar insulation in presence of different shapes, locations and sizes of cavities. AIP Conf Proc. https://doi.org/10.1063/1.5133927
8. Ilhan S, Ozdemir A, Ismalloglu H (2015) Impacts of Corona rings on the insulation performance of composite polymer insulator strings. IEEE Trans Dielect Electr Insul 2(3):1605–1612. https://doi.org/10.1109/TDEI.2015.7116356
9. Akbari E, Mirzaie M, Rahimnejad A et al (2012) Finite element analysis of disc insulator type and Corona ring effect on electric field distribution over 230-kv insulator strings. Int J Eng Technol 1(4):407–419
10. Haddad A, Warne D (2004) Advances in high voltage engineering. In: IET power and energy series 40. The Institution of Engineering and Technology, pp 1–669

Ultra-Compact All-Optical NAND Logic Gates Based on 4 × 4 MMI Coupler Using Silicon Hybrid Plasmonic Waveguides

Thi Hong Loan Nguyen, Duy Tien Le, Anh Tuan Nguyen, and Trung Thanh Le

Abstract We present a new structure based on ultra-compact cascaded multi-mode interference (CS-MMI) structures using silicon hybrid plasmonic waveguides (HPWG) for an all-optical NAND logic gate. An ultra-compact footprint of 3 × 16 μm^2 can be achieved. We show that the contrast ratios for logic 1 and logic 0 for NAND gates are about 25 dB for a ultra-high bandwidth of 70 nm, respectively. A large fabrication tolerance can be achieved by using this structure. This all-optical NAND gate can be useful for all-optical high speed networks.

1 Introduction

In recent years, the plasmonic waveguide is a kind of novel structure enabling nanoscale optical confinement. It is suitable for ultra-compact devices. One of the most important plasmonic waveguides is silicon hybrid plasmonic waveguides (HPWGs), which is compatible with SOI (silicon-on-insulator) technology, but still has a strong field [1]. The propagation loss of such waveguide is also quite small, is at order of 0.01 dB/μm, which is much lower than those traditional metal nanoplasmonic waveguides. In particular, on-chip photonic devices assisted locally by HPWGs can have acceptably low losses and ultra-compact footprints. As a result, a variety of HPWGs with modified structures have been proposed for realizing ultra-high compact devices, which is useful for photonic integrated circuits [2].

All-optical logic gates have many possible applications in optical signal processing systems and optical switching networks such as header recognition, parity checkers, and encryption systems [3]. In all-optical networks, there is a great need for implementing all-optical logic gates having small size, low power consumption and high-speed [4]. These requirements can be met by using photonic integrated

T. H. L. Nguyen
Hanoi University of Natural Resources and Environment (HUNRE), Hanoi, Vietnam

D. T. Le · A. T. Nguyen · T. T. Le (✉)
International School (VNU-IS), Vietnam National University (VNU), Hanoi, Vietnam
e-mail: thanh.le@vnu.edu.vn

Z. Zakaria and S. S. Emamian (eds.), *Recent Advances in Electrical and Electronic Engineering and Computer Science*, Lecture Notes in Electrical Engineering 865, https://doi.org/10.1007/978-981-16-9781-4_8

circuits, especially silicon photonics. There have been some approaches to realize all-optical logic gates such as Mach–Zehnder interferometer with a nonlinear phase shifter [5, 6], semiconductor optical amplifiers (SOAs) [7], microelectromechanical systems (MEMS) [8], MMI based photonic crystal [9], periodic waveguide [10], plasmonic waveguide [11] and multimode interference waveguide [12]. However, these methods require high power and/or complicated fabrication. Over the last few years, we have presented a general theory for implementing optical signal processing based on MMI elements [13, 14]. For further development, we have proposed 2 × 2, 3 × 3 and 5 × 5 MMI based structures for implementing many optical logic gates including NAND, OR, AND, NOT, XNOR, NOR gates [15].

In this paper, we present a new scheme for realizing an all-optical logic NAND gate structure based on only one 4 × 4 MMI cascaded with a 2 × 2 MMI coupler on HPWGs. The device has the advantage of ease of fabrication, large fabrication tolerance, high contrast ratio and compatibility with system-on-a- chip configuration.

2 Theory of NAND Based on Cascaded MMIs on HPWG

Figure 1 shows a structure for realizing all-optical NAND gate based on cascaded MMI couplers. The silicon hybrid plasmonic waveguide is shown in Fig. 1b. For a low loss and compact size, we choose the layer thicknesses as follows: $h_{Si} = 230$ nm,

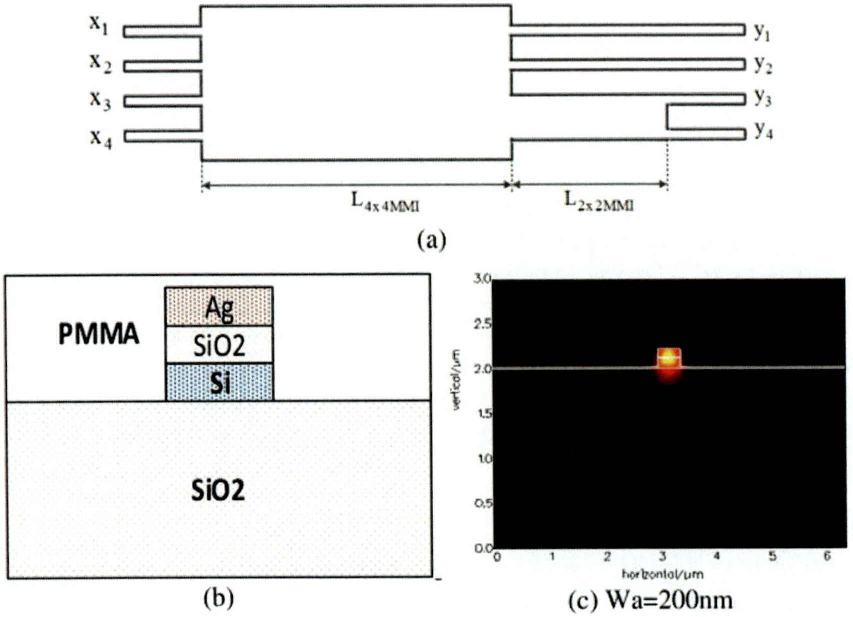

Fig. 1 **a** Proposed scheme for optical logic gates, **b** HPWG cross-sectional view and **c** Field profile of the waveguide

$h_{SiO2} = 50$ nm and $h_{Ag} = 100$ nm. It is noted that we choose the suitable thickness of SiO_2 layer working as a slot region between Ag and Si layers to balance the loss and confinement factors. The refractive indices of silicon, SiO_2, and silver are $n_{Si} = 3.455$, $n_{SiO_2} = 1.445$ and $n_{Ag} = 0.1453 + j11.3587$ at around operation wavelength 1550 nm [16]. PMMA is chosen to cover the cladding with its refractive index of 1.481.

The working principle of an MMI coupler is based on the self-imaging principle [17]. When the access waveguides with an identical width of W_a at the positions $p_i = \left(i + \frac{1}{2}\right)\frac{W_{MMI}}{N}$, the electrical field inside the MMI coupler can be expressed by [18]

$$E(p,z) = \exp(-jkz)\sum_{m=1}^{M} E_m \exp\left(j\frac{m^2\pi}{4\Lambda}z\right)\sin\left(\frac{m\pi}{W_{MMI}}p\right) \tag{1}$$

The characteristic matrices of a 4 × 4 and 2 × 2 MMIs can be expressed by [19]:

$$M_{4\times4} = \frac{1}{2}\begin{bmatrix} -1 & -\exp\left(j\frac{3\pi}{4}\right) & \exp\left(j\frac{3\pi}{4}\right) & -1 \\ \exp\left(j\frac{3\pi}{4}\right) & -1 & -1 & -\exp\left(j\frac{3\pi}{4}\right) \\ -\exp\left(j\frac{3\pi}{4}\right) & -1 & -1 & \exp\left(j\frac{3\pi}{4}\right) \\ -1 & -\exp\left(j\frac{3\pi}{4}\right) & \exp\left(j\frac{3\pi}{4}\right) & -1 \end{bmatrix} \tag{2}$$

$$M_{2\times2} = \frac{1}{\sqrt{2}}\begin{bmatrix} 1 & -j \\ -j & 1 \end{bmatrix} \tag{3}$$

where $i = 1 \div N$, and N is the number of ports of MMI coupler. For optical logic gates based on the MMI principle, information is encoded at the input and the output in amplitude or in phase. In this work, phase encoding of information is used for input signals and amplitude encoding is used for output signals. We use the logic "1" is represented by $1e^{j0}$ and logic "0" is represented by $0e^{j0}$.

To determine the logic level at the output of the device, the power in the output waveguide needs to be compared to a threshold value. This can be done electronically by connecting output ports to a photo-detector and a decision circuit. Another approach is to use an optical threshold device based on active MMI couplers instead of using an electronic threshold device [20]. Figure 1a shows the proposed scheme for optical logic gate implementation based on 4 × 4 and 2 × 2 MMI structures. By properly choosing the positions of input and output waveguides, the complex amplitude at output port y_2 can be expressed by:

$$y_2 = 0.5(jx_1 + x_2 + jx_3 - x_4) = 0.5(jx_1 - x_4) + 0.5(x_2 + jx_3) = f(x_2, x_3) \tag{4}$$

where x_1, x_4 are local oscillators and x_2, x_3 are input logic variables and y_2 is the output logic variable. Here we assume that the wavelength and polarization of the local oscillation signals and information signals are the same.

The widths of the HPWG single and multimode waveguides are chosen carefully to have a quality self-images at the output, but still guarantee the loss low enough. Due to the presence of the metal Ag in the hybrid plasmonic waveguide structure, the loss factor increases when the length is too long. We undertake the simulations to find the effective index for TE and TM modes as shown in Fig. 2. The simulations show that the widths of 2 × 2 and 4 × 4 MMI are chosen to be 700 nm and 2000 nm, respectively. Here the access waveguide width for single mode operation is Wa = 200 nm.

We use the local oscillation inputs $x_1 = 1e^{j\pi/2}$ and $x_4 = 1e^{j\pi/2}$. For input ports x_2 and x_3, 0-phase corresponds to logic 0 and π-phase corresponds to logic 1. The truth table for the NAND gate is shown in Table 1.

In order to achieve the required phase shift at the input waveguides of the MMI structure, one 1 × 1 MMI coupler is used. For an 1 × 1 MMI, the field at distance z along the multimode section can be written as [19]:

$$\psi(y,z = L_{1 \times 1MMI}) = e^{-j\beta_{0M}z}\psi(y,z = 0) \quad (5)$$

The difference of the relative phase between two arms of the multimode waveguide and single waveguide is $\Delta\varphi = (\beta_{0M} - \beta_0)L_M$, where β_0 and β_{0M} are the propagation constants of the fundamental modes of the single and multimode sections. Figure 3

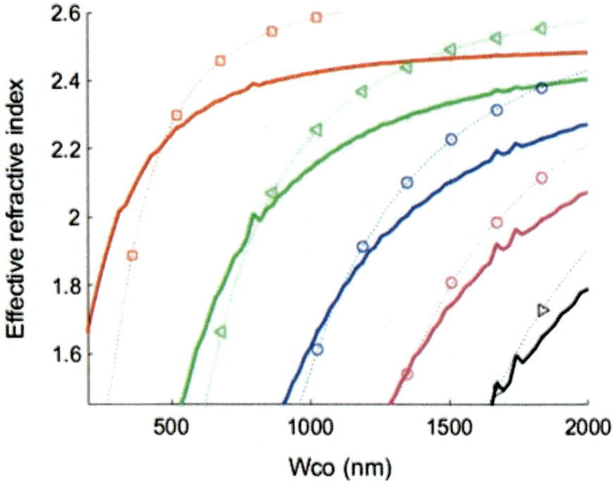

Fig. 2 Effective refractive index of the HPWG waveguide at different widths

Table 1 Truth table of the NAND logic gate

Input logic		Output logic
x_2 (phase)	x_3 (phase)	$y2 = fx_2, x_3)$
0 (0)	0 (0)	1
0 (0)	1 (π)	1
1 (π)	0 (0)	1
1 (π)	1 (π)	0

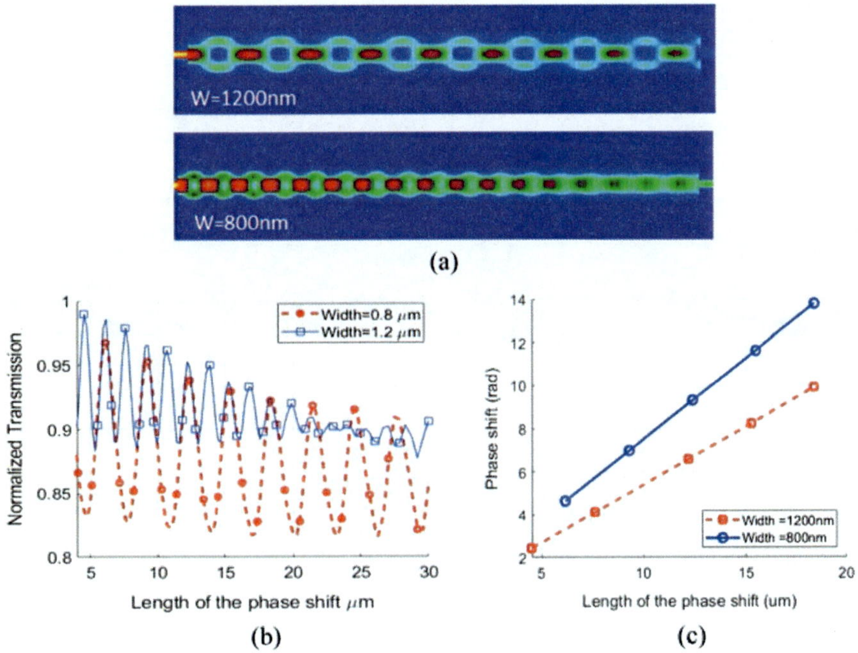

Fig. 3 Field propagation for different widths

shows the simulations of the phase shift design using an 1×1 MMI coupler with a width of 1200 and 800 nm. Figure 1a shows the field propagation and the self-image positions at different lengths are shown in Fig. 1b. Figure 1c shows the phase shift obtained for both cases.

3 Simulation Results and Discussions

In this section, light propagation through the logic gates is investigated. The numerical methods are used for the simulations. Figure 2 shows the field distributions of the NAND logic gate at wavelength of 1550 nm for input logic values of 00, 01, 10 and 11, respectively. The simulations show that there is good agreement with the theoretical analysis given by Table 1 and Fig. 4.

Figure 5 shows the normalized output powers of the NAND gate for bit 1 and 0 at different wavelengths, respectively. We can see that the output powers are quite constant over a large range of wavelength (about 70 nm).

We evaluate the performance of the optical logic gates by using the contrast ratio (CR). CR of the optical NAND gate is expressed by

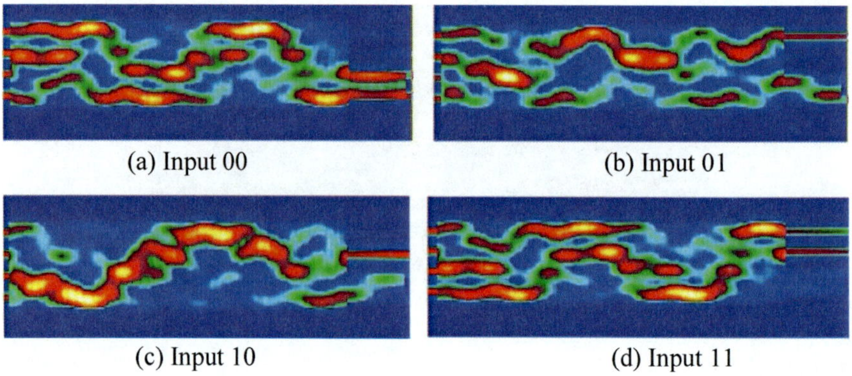

Fig. 4 NAND gate with input signals 00, 01, 10, 11

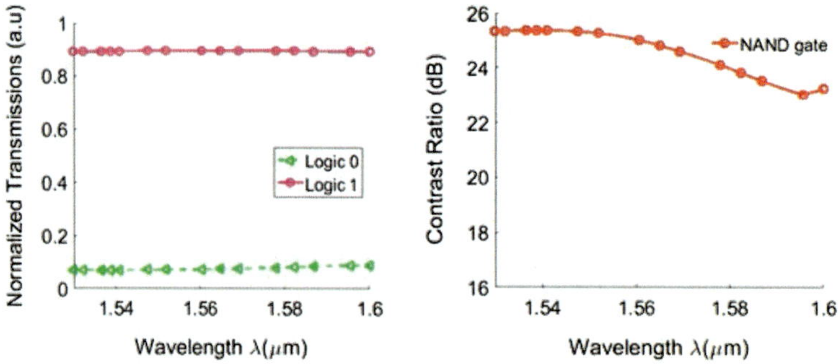

Fig. 5 **a** Normalized output powers for logic 1 and 0 and **b** contrast ratio of the NAND gate

$$CR = 10\log_{10}\left(\frac{P_{logic1}}{P_{logic0}}\right)(dB) \quad (6)$$

As a result, the CR for the NAND gates are shown in Fig. 5a. It is shown that for a bandwidth of 70 nm from 1530 to 1600 nm, the CR varies from 13 to 25.8 dB.

4 Conclusions

We have presented a structure for implementing an all-optical NAND logic gate based on silicon hybrid plasmonic waveguide. The proposed structure is based on only one 4×4 MMI cascaded with a 2×2 MMI coupler and it has advantages of ease of fabrication, large fabrication tolerance, quite large contrast ratio and high bandwidth. This new structure can be useful for optical label swapping and recognition in optical packet switching networks or on-chip signal processing applications.

Acknowledgements This research is funded by Vietnam National University, Hanoi (VNU) under project number QG.19.58.

References

1. Guan X et al (2013) Ultracompact and broadband polarization beam splitter utilizing the evanescent coupling between a hybrid plasmonic waveguide and a silicon nanowire. Opt Lett 38(16):3005–3008
2. Li C et al (2021) Subwavelength silicon photonics for on-chip mode-manipulation. PhotoniX 2(1):11
3. Le T-T (2010) Multimode interference structures for photonic signal processing. LAP Lambert Academic Publishing
4. Hardy J, Shamir J (2007) Optics inspired logic architecture. Opt Exp 15(1):150–165
5. Pramono YH et al (2000) Optical logic OR, AND, NOT and NOR gates in waveguides consisting of nonlinear material. IEICE Trans Electron **E83-C**, 1755
6. Islam MN (1989) Ultrafast all-optical logic gates based on soliton trapping in fibers. Opt Lett **14**(22), 1257–1259
7. Connelly MJ (2002) Semiconductor optical amplifiers. Springer
8. Lin LY, Goldstein EL, Tkach RW (1998) Free-space micromachined optical switches with submilli-second switching time for large-scale optical crossconnects. IEEE Photon Technol Lett 10(4):525–527
9. Hussein HME, Ali TA, Rafat NH (2018) New designs of a complete set of photonic crystals logic gates. Opt Commun 411:175–181
10. Zeng S et al (2010) Ultrasmall optical logic gates based on silicon periodic dielectric waveguides. Photon Nanostruct Fundam Appl 8(1):32–37
11. Ota M et al (2016) Plasmonic-multimode-interference-based logic circuit with simple phase adjustment. Sci Rep 6:24546
12. Li Z, Chen Z, Li B (2005) Optical pulse controlled all-optical logic gates in SiGe/Si multimode interference. Opt Exp 13(3):1033–1038
13. Cahill LW, Le TT (2008) Photonic signal processing using mmi elements. In: 10th international conference on transparent optical networks (ICTON 2008), Athens, Greece
14. Le TT, Cahill L (2011) All-optical signal processing circuits using silicon waveguides. In: The 7th international conference on broadband communications and biomedical applications, Melbourne, Australia
15. Le D, Le T, Cahill LW (2018) Optical signal processing based on 4×4 multimode interference structures. In: 2018 20th international conference on transparent optical networks (ICTON)
16. Palik ED (1985) Handbook of optical constants of solids. Academic Press
17. Bachmann M, Besse PA, Melchior H (1994) General self-imaging properties in N x N multimode interference couplers including phase relations. Appl Opt **33**(18), 3905–3911
18. Heaton JM, Jenkins RM (1999) General matrix theory of self-imaging in multimode interference (MMI) couplers. IEEE Photon Technol Lett 11(2):212–214
19. Le T-T (2010) Multimode interference structures for photonic signal processing: modeling and design. Lambert Academic Publishing, Germany. ISBN 3838361199
20. Rodgers JS, Ralph SE, Kenan RP (2000) Self guiding multimode interference threshold switch. Opt Lett **25**(23)

Microstrip Patch Antenna Arrays Design for 5G Wireless Backhaul Application at 3.5 GHz

Ahmed Jamal Abdullah Al-Gburi, Zahriladha Zakaria, Imran Mohd Ibrahim, and Elzameera Bt Abdul Halim

Abstract Fifth-generation wireless (5G) is the next cellular technology iteration, designed to significantly improve wireless network speed and responsiveness. Besides, 5G allows a substantial increase in the number of facts sent through the wireless system owing to greater bandwidth and advanced antenna technology. However, in this paper, hexagonal microstrip patch antennas are designed and simulated for wireless backhaul at 3.5 GHz. Four types of antennas were simulated using Computer Simulation Technology (CST) software. The developed process started from designing the single element up to 1 × 8 arrays elements. The proposed 1 × 8 arrays antenna feds by microstrip corporate feed line and provided directional radiation, which is useful for the base station to provide high quality and high capacity network connectivity. Moreover, this type of antenna is focused on long-distance point to point connections. The finalised antenna offered a high gain of 6.938 dB at 3.5 GHz and a return loss of -10 dB. Overall antenna performance confirms that the proposed hexagonal patch antenna is proper for 5G communication.

Keywords 5G · Array antenna · Hexagon shaped slotted antenna · Microstrip antenna

1 Introduction

The fifth-generation (5G) of communication has been widely discussed as a way to provide high-speed connectivity in the future. The design and validation of the 5G communication system depend upon understanding the propagation channels [1], and a considerable body channel estimate is needed. At the moment, 5G mobile systems are expanding their coverage to accommodate higher data rates. For example, in 2015the, World Radio Communication (WRC) allocated frequency bands below

A. J. A. Al-Gburi · Z. Zakaria (✉) · I. M. Ibrahim · E. B. A. Halim
Centre for Telecommunication Research & Innovation (CeTRI), Fakulti Kejuruteraan Elektronik Dan Kejuruteraan Komputer (FKEKK), Universiti Teknikal Malaysia Melaka (UTeM), Hang Tuah JayaDurian Tunggal, 76100 Melaka, Malaysia
e-mail: zahriladha@utem.edu.my

Z. Zakaria and S. S. Emamian (eds.), *Recent Advances in Electrical and Electronic Engineering and Computer Science*, Lecture Notes in Electrical Engineering 865, https://doi.org/10.1007/978-981-16-9781-4_9

6 GHz were widely discussed. The following frequency ranges were proposed: 470–694 MHz, 1427–1518 MHz, 3300–3800 MHz, and 4500–4990 MHz. Among them, 3.5 GHz has been generally thought of, as it tends to be acknowledged for the more significant part of nations [2, 3].

The principle preferences of 5G are data rates can be reached and accomplished around 10 Gbps, which can give a superior experience to the client and speeds up for download and upload. Other than that, 5G can improve the resolution and bi-directional extensive bandwidth shaping [4]. Moreover, 5G mm-wave is ready to accomplish inactivity rate under 1 ms, which can bring about a quick association foundation and delivery with the 5G organisation by 5G cell phones, and traffic load is diminished. In addition, 5G is conceivable to give a uniform, steady and continuous network across the world. 5G innovation is in a situation to accumulate all networks on one platform. It offers a 10 × decrease in latency, 100 × traffic limit, 10 × connection density, 3 × spectrum efficiency and 100 × network proficiency. 5th generation is all the more effectively to oversee contrasted with the past ages.

Wireless backhaul is used to encourage data in critical networks, such as a web or a selected organisation of a vast industry, university institution or government association, from an end-customer to a centre of a significant network [5, 6]. The term can likewise allude to the transmission of organising data over an elective wireless course when the ordinary course is unavailable or overtaxed. The regular principal procedure of wireless backhaul incorporates a microwave system despite satellites being used.

Few studies were conducted by reviewing the works of literature on the design topic of array antennas for wireless backhaul 3.5 GHz 5G applications. The following is a summary of reviews. Reference [7] designed a bow-tie antenna with high gain for next-generation base station applications operating between 2.5 and 3.9 GHz. It used substrate Rogers 5880 with the thickness of $h = 1.575$ mm, the relative permittivity of 2.2 and loss tangent of 0.0009. The assessment of the peak gain of the bow-tie antenna with and without the EECSR array was performed, and it shows the gain increases from 9.7 dB at 3.4 GHz to 10.9 dB at 3.6 GHz. In addition, the antenna reflection coefficient is better than −10 dB in the 2.5–3.9 GHz WiMAX band. In [8] designed a new Bluetooth bow-tie antenna array, worldwide microwave access (WiMAX) and wireless local area network (WLAN) applications. This antenna operates in three states of the frequency band which is 2.27–2.68, 3.25–3.92 and 4.14–6 GHz. The radiation pattern and peak gains were measured in an anechoic chamber lab at three frequencies, 2.4, 3.5, and 5.5 GHz. The gain obtained at 2.4, 3.5, and 5.5 GHz is 4.37, 6.1 and 9.8, respectively. In [9], proposed microstrip rectangular patch antenna with centre frequency at 2.5 GHz for WiMAX application. FR-4 has been used as a substrate for this paper with a dielectric constant of 4.9 and thickness of 1.6 mm. The return loss simulation shows −20.24 dB at 2.5 GHz and −22.22 dB return loss at 2.67 GHz for measurement results. The simulated result of the array antenna design creates more intensity or centre at the focal point of the radiation with the directivity and gain of 10.25 and 5.732 dB, which is better than a single patch. In [10], the authors proposed a U-slot microstrip antenna design on a resonant frequency of 3.6 GHz. The 1.6 mm thick RT/Duriod 5880 material and 2.2 dielectric constant with loss tangent 0.0009 was utilised. The result for return loss of antenna was −26 dB at the resonant frequency 3.6 GHz.

In this article, a microstrip planar array antenna has been performed for 5G communication practices. The suggested antenna contributes the benefits of low transmission loss, with a high antenna gain of 6.938 dB at 3.5 GHz. The finalised array antenna consists of 8 hexagonal parts together placed at the substrate array patch. The FR4 substrate was used for the design process. The simulated outcomes prove that the introduced antenna is a good candidate for 5G scenarios.

2 Design Antenna and Methodology

The single patch at the necessary frequency was initially calculated in order to construct this strategy. CST 2019 (Computer Simulation Technology) was utilised to simulate the design. The antenna's fundamental features, such as the resonance frequency, return loss, bandwidth, gain and directivity, are all factors that need to be considered when designing the antenna. The process designing of the antenna array has been completed.

2.1 Design Antenna Specifications

The specification is indicated for prototype evolution and was established by constructing on a frequency of 3.5 GHz, as suggested in [11, 12]. Table 1 presents the antenna specifications; the selected return loss for this research is below −10 dB and that is broadly carried out by antenna design; in this design, the FR4 substrate with a dielectric material ($\varepsilon r = 4.3$) with dielectric loss tangent (tan δ) of 0.035 and the height of substrate (h) 1.6 mm were used.

Table 1 Design parameters of the proposed antenna

Antenna parameter	Value
Dielectric constant, εr	4.3
Thickness of copper	0.035 mm
Thickness of substrate	1.6 mm
Frequency, fr	3.5 GHz
Return loss, RL	<−10 dB
Gain, G	6–10 dB

2.2 Design Antenna Process

In this section, the single element of the modelled antenna integrated with elements arrays were designed. The progress starts from preparing the single hexagonal with a total dimension of 27 × 35 × 1.6 mm, sequentially employing a hexagonal patch shape of a microstrip antenna.

The hexagonal-shaped slotted antenna was designed at a 3.5 GHz resonant frequency. To determine the radiating patch elements, appropriate equations are employed but still dependent on the patch form, either circular, rectangular or triangular [13]. In this study, a rectangle patch is the basic starting design. The formulae for deciding rectangular patch dimensions are as follows [14]:

The width of the patch can be calculated using the formula below:

$$W = \frac{c}{2f_r\sqrt{\frac{\varepsilon_r+1}{2}}} \tag{1}$$

where:

εr = *material relative constant.*

For the actual length of a patch antenna, it is calculated using this formula:

$$L = \frac{C}{2f\sqrt{\varepsilon_{reff}}} - 2\Delta L \tag{2}$$

where:

$\varepsilon reff$ = effective dielectric constant, and ΔL is the length extension.

2.3 Design Parameter of Hexagonal-Shaped Slotted Antenna.

After designing the single element, proceeding with the complete elements design, the antenna design may be adjusted to meet 5G requirements using the research method. The width, length, and impact on frequency gain return loss, directivity, and gain are all measured and compared using the research method. Figure 1. illustrates the design structures for the front of the single 1 × 2, 1 × 4, and 1 × 8 array antennas. Meanwhile, Table 2 shows the optimised parameters for the 1 × 8 microstrip array antenna. The overall antenna dimensions are 200 × 160 × 1.6 mm.

The parameters of the patch parts are kept fixed. The dimension of the feed line elements is also held the same as that in the original patch design. The ground plane of the proposed hexagonal array antenna has square slots among the radiating patches. The feeding arrangement of the array is made of microstrip T-coupling power dividers with respect to λ/4 of impedance mismatch to ultimately achieve input the impedance matching of about 50 Ω at the input gate of the array.

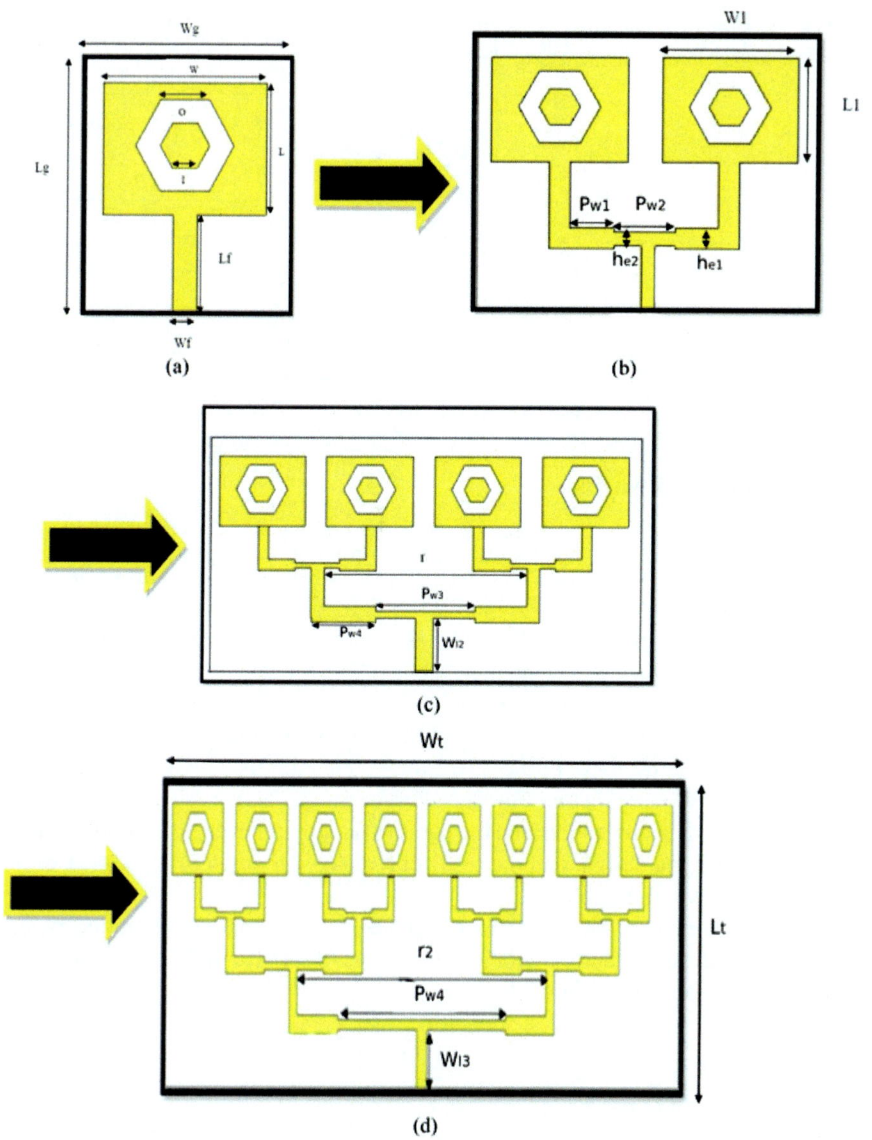

Fig. 1 Geometries of the proposed hexagonal antenna in different array elements: **a** single array element, **b** 1 × 2 array elements, **c** 1 × 4 array elements and **d** 1 × 8 array elements

Table 2 Dimensions of the modelled hexagonal antenna

Symbol	Value (mm)	Symbol	Value (mm)
W_g	26	P_{w1}	5
L_g	35	P_{w2}	7
W	20	h_{e1}	3
L	15	h_{e2}	1.7
L_f	15.8	P_{w3}	18
W_f	4	r_1	40
O	6	W_{l2}	15
I	2	r_2	97
P_{w4}	60	W_{l3}	30
W_t	200	L_t	160

3 Simulation Results

3.1 Return Loss

Referring to Fig. 2a, the result of S-Parameters or return loss of the single hexagonal-shaped slotted microstrip patch antenna is shown. The simulated resonance frequency of the antenna is 3.5 GHz, with a return loss of −2.119 dB.

After the optimisation procedures and the parametric analysis on the proposed array antennas, the response of the 1 × 2 array antenna was 3.5 GHz with a return loss of −1.567 dB, as shown in Fig. 2b.

The optimisation procedures and the parametric research were also conducted for 1 × 4 array antenna design, and the results produced a return loss of −14.11 dB at 3.5 GHz, as shown in Fig. 2c

A 1 × 8 parametric analysis was conducted for the final configuration of the array antenna design, resulting in a return loss of −12,45 dB at 3.5 GHz resonance frequency as duplicated in Fig. 2d.

3.2 Realised Gain

The gain of the antenna is the next parameter need to be considered in this project. Gain is one of the most major factors to be evaluated because the aim is to create a high gain array antenna capable of 5G applications. As a result, shown in Fig. 3, the gain obtained for single, 1 × 2, 1 × 4 and 1 × 8 antennas are 0.404 dB, 1.396 dB, 5.366 dB and 6.938 dB, respectively. The 1 × 8 array antenna with a gain of 6.942 dB proves that the antenna can cover the more extensive distance concerning 5G applications.

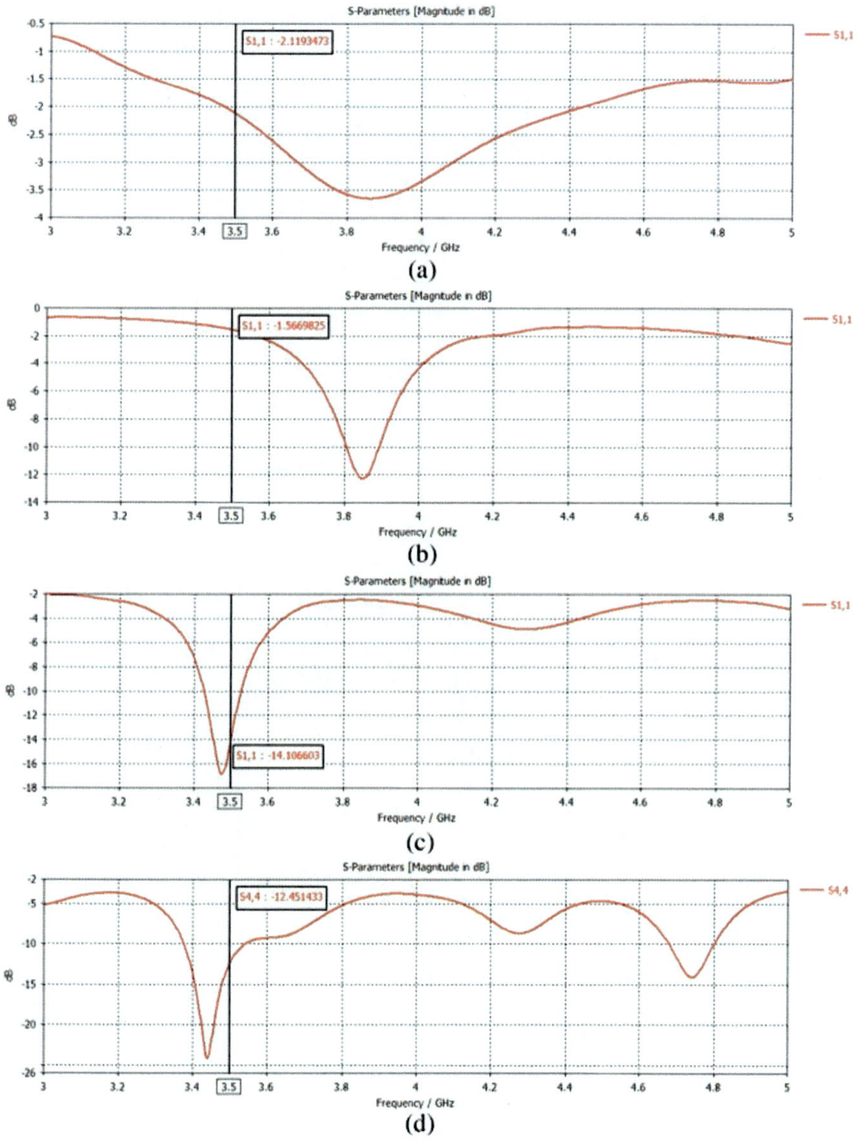

Fig. 2 Return loss performances for **a** single antenna, **b** 1 × 2 array antenna, **c** 1 × 4 array antenna, and **d** 1 × 8 array antenna

3.3 Radiation Pattern

The result was achieved by the simulation given in terms of radiation pattern for the four configuration designs, which are the single antenna design, 1 × 2, 1 × 4, and

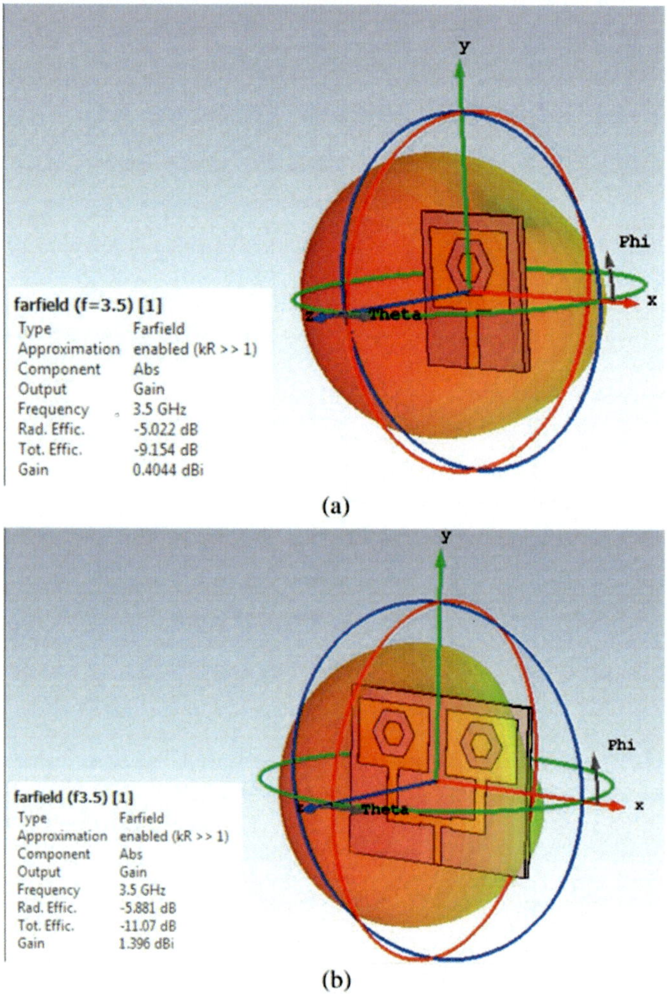

Fig. 3 Realised gain for **a** single antenna, **b** 1 × 2 array antenna, **c** 1 × 4 array antenna, and **d** 1 × 8 array antenna

1 × 8 array antennas. These array antenna designs are directional, which implies that the radiation is concentrated in one direction rather than radiating in all directions, as illustrated in Fig. 4. According to the figure below, the main peak beam of radiation pattern (blue colour) is close to 0°, demonstrating that it has a substantial array as well as good radiating characteristics at their operating bands, thus suitable for 5G application. In addition, the smaller the lobe, the stronger the radiation.

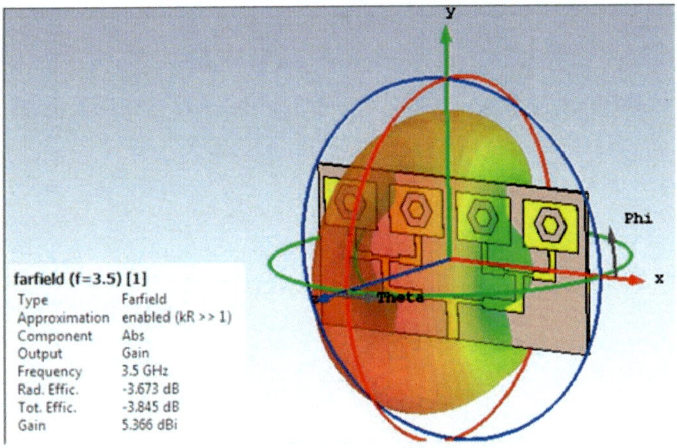

(c)

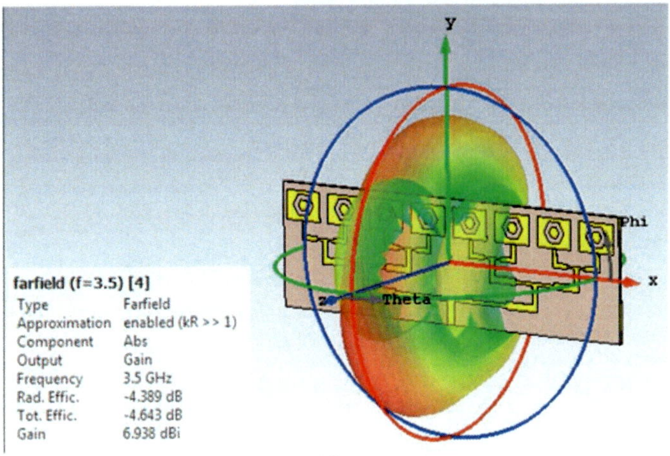

(d)

Fig. 3 (continued)

4 Comparison Between the Array Antennas

This analysis focuses on identifying the optimal design for this research. In this study, all antenna designs were separated and compared with each feature suitable for analysis, which was return loss, gain, and directivity. Table 3 shows the findings for comparing various antenna designs given in this chapter. Figure 5 shows the difference between return loss results of all antenna designs.

A comprehensive presentation is carried out in this stage, and it's presented next. First, the discussion focused on each antenna's return loss. The 1 × 4 hexagonal shaped slotted microstrip patch antenna shows the lowest return loss of −14.11 dB followed by 1 × 8 array antenna, single hexagonal-shaped slotted microstrip patch

Fig. 4 Radiation pattern results in **a** single antenna, **b** 1 × 2 array antenna, **c** 1 × 4 array antenna, and **d** 1 × 8 array antenna

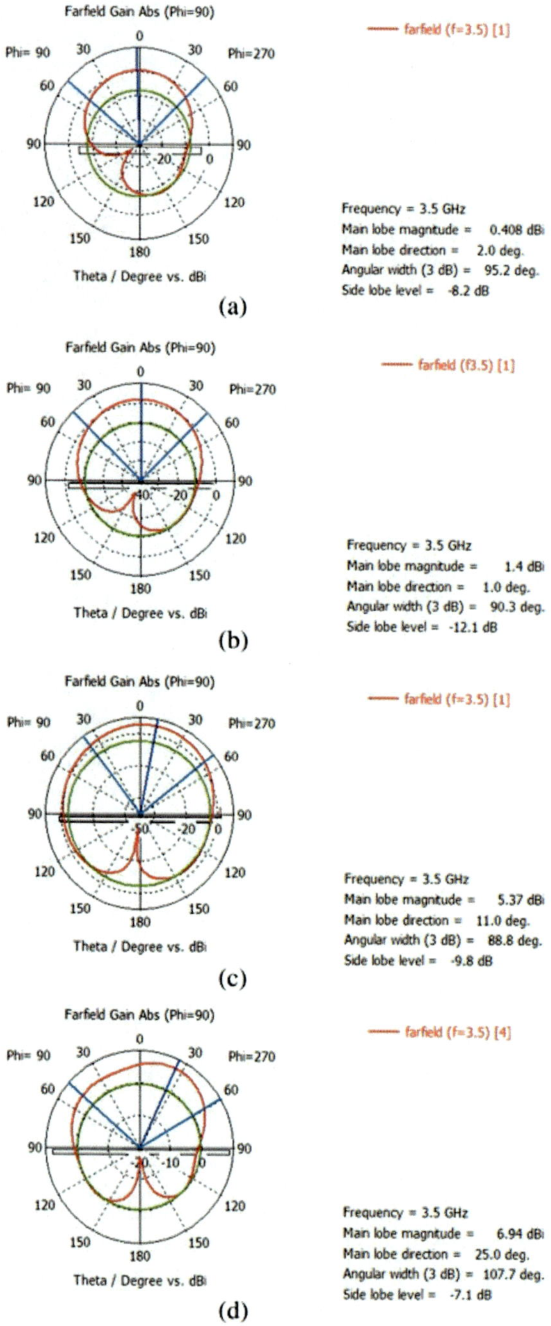

Table 3 Comparison with various designs of antennas at 3.5 GHz

Antenna type	Return loss (dB)	Gain (dB)	Directivity (dB)
Single hexagonal shaped slotted microstrip patch antenna	−2.12	0.404	5.43
1 X 2 hexagonal shaped slotted microstrip patch antenna	−1.57	1.396	7.28
1 X 4 hexagonal shaped slotted microstrip patch antenna	−14.11	5.366	9.04
1 X 8 hexagonal shaped slotted microstrip patch antenna	−12.45	6.938	11.33

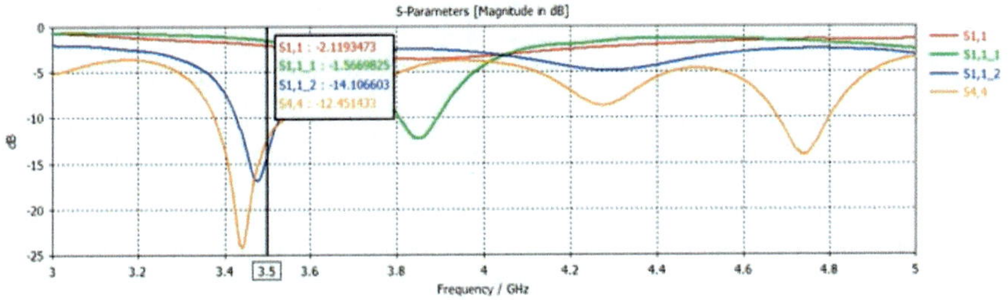

Fig. 5 Return loss performance of all antennas

antenna and 1 × 2 array antenna with a return loss of −12.45 dB, −2.12 dB and −1.57 dB correspondingly. The result showed a minimal return loss for the antenna as this paper aims to achieve at least −10 dB.

The next parameter evaluated is an antenna gain. Since constructing a high-gain antenna is the main objective of this research, this research focused on identifying an antenna capable of generating a high gain. From Table 3 above, 1 × 8 array antenna had the highest gain of 6.938 dB compared to 5.366 dB, 1.396 dB and 0.404 dB antenna.

As for the result, the single antenna simulation yielded a directivity of 5.43 dB. Meanwhile, the simulation for 1 × 2 array antenna is 7.28 dB, while the simulation for 1 × 4 and 1 × 8 array antenna is 9.04 and 11.33 dB, respectively. The gain always corresponds to the directivity. The directivity increases as the gain increases. The greater the directivity, the stronger the signal transmitted via the aperture of the array antenna.

5 Conclusion

In this paper, the hexagonal shaped slotted array antenna for 5G applications is suggested. For each design, the gain was improved using the slotted approach. At 3.5 GHz, 1 × 8 array antenna obtained, the simulated S11 was −12.45 dB, whereas the gain was 6.938 dB. Due to the enhanced performance, the suggested antenna might be a good option for 5th Generation Wireless Systems (5G), which demand a high gain and a return loss of less than −10 dB.

Acknowledgements We would like to thanks to Ministry of Higher Education and UTeM through FRGS Grant F00430 FRGS/1/2020/FKEKK-CETRI/F00430 that support this research.

References

1. Imperatore P, Salvadori E, Chlamtac I (2007) Path loss measurements at 3.5 GHz: A trial test WiMAX based in rural environment. In: Proceedings of the 3rd international conference on testbeds and research infrastructures for the development of networks and communities, Lake Buena Vista, FL, USA
2. Anguera J, Puente C, Borja C et al (2003) Dual-frequency broad-band stacked microstrip patch antenna. IEEE Antennas Wirel Propag Lett 2:36–39
3. Zeain MY, Zakaria Z, Abu M et al (2020) Design of helical antenna for next generation wireless communication. Prz. Elektrotechniczny 11:96–99
4. Kamaruddin RAA Ibrahim IBM, Al-Gburi AJA et al. (2021) Return loss improvement of radial line slot array antennas on closed ring resonator structure at 28 GHz. Przegląd Elektrotechniczny 1(5), 67–71.
5. Wang N, Hossain E, Bhargava VK (2016) Joint downlink cell association and bandwidth allocation for wireless backhauling in Two-Tier HetNets with large-scale antenna arrays. IEEE Trans Wirel Commun 15(5):3251–3268
6. Abdulhameed MK, Hashim SR, Abdalhameed NK et al (2022) Increasing radiation power in half width microstrip leaky wave antenna by using slots technique. Int J Electr Comput Eng 12(1):392–398
7. Dadgarpour A, Zarghooni B, Virdee BS et al (2015) High-gain end-fire bow-tie antenna using artificial dielectric layers. IET Microwav. Antennas Propag 9(12):1254–1259
8. Applications W (2015) Frequency-Recon fi gurable Bow-Tie Antenna for. IEEE Antennas Wirel Propag Lett 14:171–174
9. Wahab, NA, Bin Maslan Z, Muhamad WNW et al (2010) Microstrip rectangular 4×1 patch array antenna at 2.5GHz for WiMax application. In: Proceedings-2nd International Conference on Computational Intelligence, Communication Systems and Networks, Liverpool, UK, 2010, pp 164–168
10. A. Bhargava and P. Sinha, "Multi Rectangular Slotted Hexa Band Micro-strip Patch Antenna for Multiple Wireless Applications," in Proceedings of the *3rd International Conference on Communication and Electronics Systems, ICCES 2018*, Coimbatore, India, pp. 902–904.
11. Abdulhameed MK, Kod MS, Al-gburi AJA (2021) Enhancement of elevation angle for an array leaky-wave antenna. Prz Elektrotechniczny 8:109–113
12. Al-Gburi AJA, Ibrahim I, Zakaria Z et al (2021) High gain of UWB CPW-fed Mercedes-shaped printed monopole antennas for UWB applications. Prz Elektrotechniczny 5:70–73
13. Al-Gburi AJA, Ibrahim I, Zakaria Z et al (2019) Bandwidth and gain enhancement of ultra-wideband monopole antenna using MEBG structure. ARPN J Eng Appl Sci 14(10), 3390–3393
14. Keriee HH, Rahim MKA, Nayyef NA et al (2020) High gain antenna at 915 MHz for off grid wireless networks. Bull Electr Eng Inf 9(6):2449–2454

Transformer Life Estimation Based on Degree of Polymerization with GUI

N. I. Hassan, M. A. Othman, and W. J. Gim

Abstract Power transformer (PT) is one of the important devices in power transmission system. The end-of-life indication can be observed as failure of solid insulation and insulating paper of the PT. The degree of polymerization (DP) is widely used measurement to examine the health state of the PT insulating paper. Currently, DP value of the insulating paper in the PT can be estimated by multiple mathematical models using one of measured Furan compounds, 2-furaldehyde (2-FAL). A formula was used to estimate the health state of the insulating papers in 274 units of PTs with 2-FAL data that operate in Selangor, Malaysia. As a result, the end-of-life for the 274 units transformer are analysed and showed in Graphical User Interface (GUI).

1 Introduction

A power transformer (PT) is an essential electrical device used in electricity transmission and distribution system that can regulate the voltage level of electricity for transmission purposes [1]. In that case, PT asset management becomes an important aspect because failure of the PT not only requires for new replacement that is costly but also gives several impacts such as high risk igniting the PT due to the direct contact of the high voltage components of PT with the insulating oil [1–5]. Theoretically, the life expectancy of PT is related to insulating properties, thus inspecting the samples of the insulating paper can be used to investigate the health state of the PT. Degree of polymerization or DP is the average number of anhydroglucose rings present in cellulosic macromolecule chain of insulating cellulosic paper. DP is a well-known measurement for the degradation of the insulating paper in PT. The insulating paper in new and healthy state has an approximate DP value of 1200 while the insulating paper in its deteriorated state has the approximate DP value of 200 according to

N. I. Hassan (✉) · W. J. Gim
Heriot-Watt University Malaysia, Putrajaya, Malaysia
e-mail: n.hassan@hw.ac.uk

M. A. Othman
Universiti Teknikal Malaysia Melaka, Melaka, Malaysia

Z. Zakaria and S. S. Emamian (eds.), *Recent Advances in Electrical and Electronic Engineering and Computer Science*, Lecture Notes in Electrical Engineering 865, https://doi.org/10.1007/978-981-16-9781-4_10

IEEE C57.140 standard and thus it can be aided in the life estimation of the PT [6, 7]. Dissolved gas analysis (DGA) and Furan compounds analysis are the common tests conducted on the insulating oil. Therefore, Furan compounds acquired from the insulating oil are considered related to DP value of the insulating paper in the PT and 2-furaldehyde (2-FAL) is considered the main element of the Furan compounds [2–4, 6]. 2-FAL measured from the PT is correlated to the DP value of the insulating paper [6, 8]. Thus, 2-FAL becomes typical in life estimation of PT for the last two decades due to its direct relationship with the aging of the insulating paper in PT [6]. Graphical user interface (GUI) could be beneficial to include as it can provide convenience for the experts to perform the life estimation of PTs in Malaysia. In this paper, the life expectancy for 274 unit of PT with 2-FAL data were showed in histogram and in designed GUI.

2 Methodology

2.1 Transformer Life Estimation

In this study, 274 units of PT sample with 2-FAL measured were analysed. Microsoft Excel was used to arrange and extract the required data from the oil data and recorded down the important statistics. MATLAB programme was used to conduct accuracy assessment of the mathematical models. It was also used to build a GUI that could conduct life estimation for the PTs and visualise their deterioration rate in terms of DP values on a graph. The voltage ratings of the investigated PTs were 30/11 and 33/11 kV while their power ratings were 7.5, 15 and 30 MVA. Figure 1 shows the flowchart illustrating the process of obtaining the estimated DP values using measured 2-FAL. Three case study shown in the flowchart is not mentioned in this paper.

In the oil data provided by the experts, there were many oil parameters measured from the PTs through various types of diagnostic tests including DGA and oil quality analysis. These oil parameters consisted of dissolved gases and dielectric characteristics. Dissolved gases were produced when the insulating oil was breaking down under abnormal electrical and thermal stresses [3]. The dissolved gases are shown in Table 1.

Due to degradation in insulating oil via chemical reactions, it resulted the changes in the electrical, chemical and physical properties of the insulating oil which leads to decreasing of dielectric properties [3]. The dielectric characteristics that were measured using oil quality analysis as shown in Table 2.

Correlation analysis is a common and flexible data analytic system where it can be used to analyse the strength of relationship between two quantitative variables. In addition, linear correlation is considered to be the basic correlation analysis that can be used whenever a study of the relationship between a quantitative variable, the response variable, and a factor of interest, the predictor variable [7].

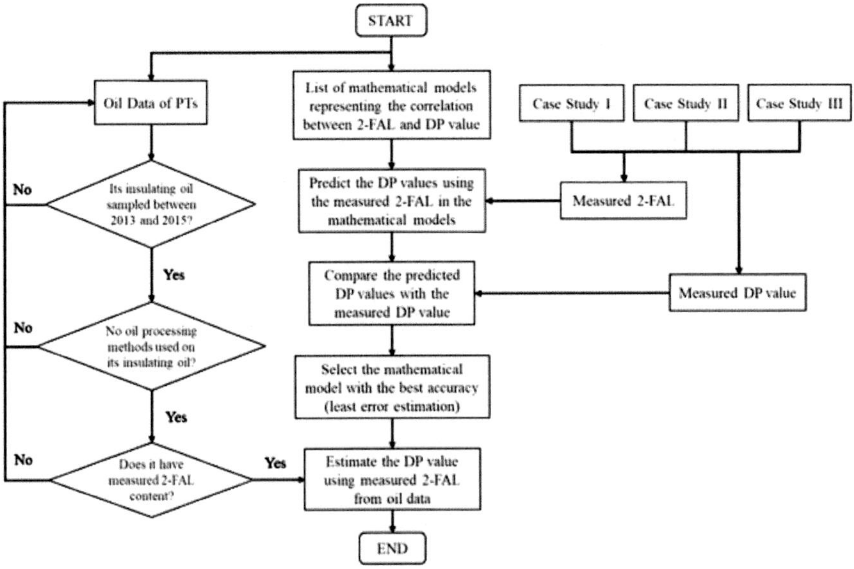

Fig. 1 Flowchart illustrating the process of obtaining the estimated DP values using measured 2-FAL from Malaysian PTs

Table 1 Dissolved gases as the potential oil parameters

Dissolved gases in ppm	
• Hydrogen (H_2) • Methane (CH_4) • Oxygen (O_2) • Acetylene (C_2H_2) • Nitrogen (N_2)	• Ethylene (C_2H_4) • Carbon monoxide (CO) • Ethane (C_2H_6) • Carbon dioxide (CO_2)

Table 2 Dielectric characteristics as the potential oil parameters

Dielectric characteristics	
• Dielectric breakdown voltage (kV) • Moisture content (ppm) • Acidity (mg KOH/g) • Oxidation inhibitor content (%) • Interfacial tension (mN/m) • Colour (Scale 0 to 8) • Relative density • Viscosity (mm^2/s)	

DP value could now be estimated using selected mathematical model. Therefore, the percentage used life of PT could be derived and is mathematically represented by Eq. 1:

$$\%used = \left(\frac{1200 - DP_{est}}{1200 - 200}\right)100\% \tag{1}$$

The used life of PT was referred to the servicing age of the PT based upon the build and sample dates and is mathematically represented by Eq. 2:

$$Used\ Life = Sample\ Date - Build\ Date \tag{2}$$

The maximum lifespan or total life of PT could be derived and is mathematically represented by Eq. 3:

$$Total\ Life = Used\ Life \cdot \frac{100\%}{\%Used} \tag{3}$$

2.2 *Graphical User Interface (GUI)*

GUI adapted the concepts of estimation for DP values to perform the life estimation of PTs. The main features for the GUI had to be the inputs for 2-FAL, the oil parameters and the inputs for dates when the PTs were built and when the samples of their insulating oil were taken. Hence, the GUI could estimate the DP values using the algorithms in the selected mathematical model and utilised the estimated DP to predict the life conditions of the PT. The displays illustrate the health condition of the insulating paper, the maximum lifespan of PTs and the expected date the PTs was going to be fully exploited Fig. 2.

3 Results and Discussions

As stated, 274 units of PTs were being analysed. Linear correlation could be applied on the potential input oil parameters among furan compounds, dissolved gases and dielectric characteristics of the 274 units of investigated PTs to determine the highly correlated and statistically significant oil parameters such that the advantages of their relationships with the DP values could be utilised. The oil parameters would require performing linear correlation and it could be carried out in Minitab software.

In Table 3, it was observed that the *p*-values for H_2, N_2 and CO_2 were less than the significance level, α, of 0.05, indicating that the probabilities of these dissolved gases having no relationship with the DP value were very low and their correlations were considered statistically significant. However, it was observed that the magnitudes of the Pearson correlation coefficients for these three dissolved gases were particularly

Table 3 Pearson correlation coefficients and *p*-value between dissolved gases and DP value for 274 units PTs

Dissolved Gases	DP Value	
	Pearson's *r*	*p*-value
H_2 (ppm)	0.1988	0.0009
O_2 (ppm)	−0.0664	0.2735
N_2 (ppm)	0.1367	0.0236
CO (ppm)	0.0042	0.9453
CO_2 (ppm)	−0.1920	0.0014

low. According to Ratnasari et al. [9], when the magnitude of size of Pearson correlation coefficient was between 0 and 0.2, the strength of the correlation between two variables was negligible.

With mathematical model proposed, the DP value of the insulating paper in any PTs could then be estimated as long as the condition for each model was fulfilled. The expected end-of-life date for the PTs that had measured 2-FAL were plotted in terms histogram in Fig. 3.

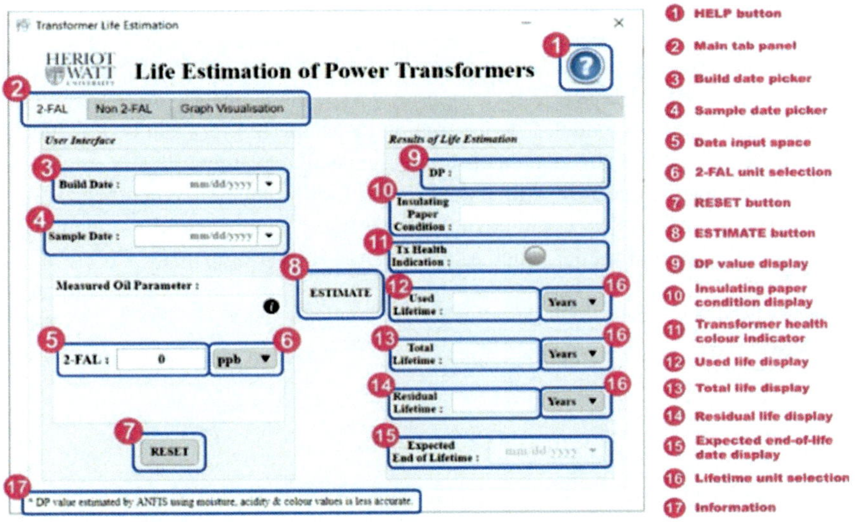

Fig. 2 Design of GUI of transformer life estimation for "2-FAL" with their features labelled

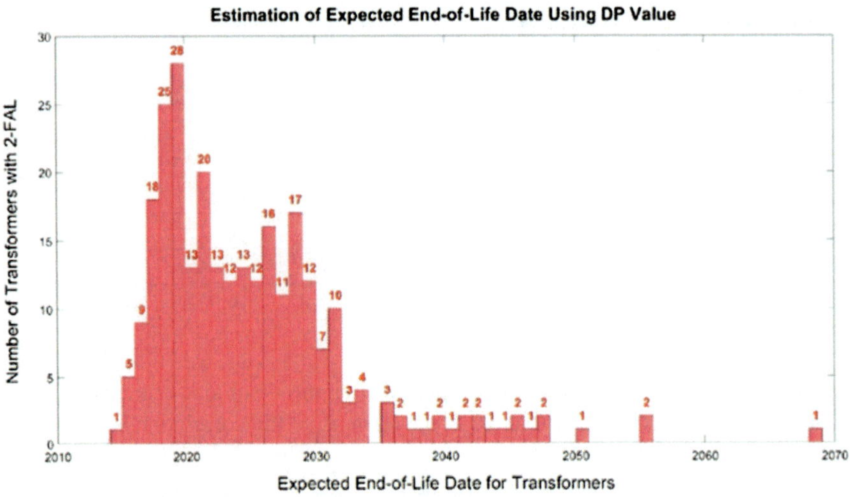

Fig. 3 Histogram displaying the expected end-of-life date for 274 units of PTs with measured 2-FAL

4 Conclusions

DP value representing the health condition of the insulating paper in 274 units PTs was estimated by applying the 2-FAL measured from the insulating oil in Malaysian PTs to a mathematical model. Life estimation of Malaysian PTs can be conducted to the expected end-of-life date with the estimated DP values. Finally, a GUI is specially designed to perform life estimation of Malaysian PTs. Nonetheless, it is just an estimation, and they could be incorrect as many factors can still influence the rate of deterioration of the PTs such as the types of insulating paper.

Acknowledgements Thank you to Universiti Teknikal Malaysia Melaka (UTeM) for the research collaboration.

References

1. Jain ST et al (2015) Transformer Failures, Causes and Impact. In: International conference data mining, civil and mechanical engineering (ICDMCME'2015)
2. Abd. El.-Aal RA et. Al (2019) Prediction of transformers conditions and lifetime using furan compounds analysis. IEEE
3. R. R. Hossain RR et al (2017) Pattern analysis of oil diagnostic results with service age of power transformers. In: IEEE Calcutta conference (CALCON)
4. Maduna TGL et al (2019) Comparative analysis of life estimation and reliability of power transformers. In: Southern African universities power engineering conference/robotics and mechatronics/Pattern recognition association of South Africa (SAUPEC/RobMech/PRASA)

5. Bhatia NK et al (2020) Machine learning-based regression and classification models for oil assessment of power transformers. In: IEEE international conference on informatics, IoT, and enabling technologies (ICIoT), Doha
6. Prasojo RA et al (2017) Transformer paper expected life estimation using ANFIS based on oil characteristics and dissolved gases (case study: Indonesian transformers). In: Energies
7. IEEE Guide for Evaluation and Reconditioning of Liquid Immersed Power Transformers. IEEE Std C57.140™ (2017)
8. Chendong X et al (1991) Monitoring paper insulation aging by measuring furfural contents in oil. In: Proceedings of 7th international symposium high voltage engineering
9. Ratnasari D et al (2016) The correlation between effective renal plasmaflow (ERPF) and glomerular filtration rate (GFR) with renalscintigraphy99mTc-DTPA study. J Phys Conf Ser

Velocity-Based Control of Piston Trajectory in a Free-Piston Linear Electric Generator by Load Current Modulation

Ahsan Bashir, Saiful A. Zulkifli, and Abd Rashid Abd Aziz

Abstract A free-piston engine (FPE) is a type of linearly reciprocating internal combustion engine (ICE) that has several advantages over typical crank-slider engines. When combined with a linear electric generator, the resultant system is a free-piston linear electric generator (FPLG), which may be used for vehicle propulsion and portable power generation. However, displacement endpoints of the piston-permanent magnet assembly are intrinsically unstable even with fixed electrical load, due to fluctuating combustion pressures. Engine stoppage can be caused by low combustion pressures and misfires, due to absence of mechanical flywheel. This work aims to achieve operational stability by controlling electrical load current and targeting a piston trajectory in each cycle. The present paper discusses results of simulation of the FPLG on *Matlab-Simulink*. Findings shed light on the effects of combustion fluctuation and control of the load current on system operation. By controlling the electrical current drain based on displacement and velocity, continuous steady-state operation can thus be achieved.

A. Bashir (✉) · S. A. Zulkifli
Department of Electrical & Electronics Engineering, Universiti Teknologi PETRONAS, Seri Iskandar, Perak, Malaysia
e-mail: ahsan_19001685@utp.edu.my

S. A. Zulkifli
e-mail: saifulazrin_mz@utp.edu.my

A. R. A. Aziz
Department of Mechanical Engineering, Universiti Teknologi PETRONAS, Seri Iskandar, Perak, Malaysia

S. A. Zulkifli · A. R. A. Aziz
Centre for Automotive Research & Electric Mobility, Universiti Teknologi PETRONAS, Seri Iskandar, Perak, Malaysia

Z. Zakaria and S. S. Emamian (eds.), *Recent Advances in Electrical and Electronic Engineering and Computer Science*, Lecture Notes in Electrical Engineering 865, https://doi.org/10.1007/978-981-16-9781-4_11

1 Introduction

The evolvement of alternative energy technologies is crucial for sustainable transportation [1–3]. Linearly-reciprocating motion of a single-moving assembly comprising of one or more pistons linked to a linear shaft produces electricity directly in an FPLG (Fig. 1). Engine friction caused by numerous components in conventional engine and piston friction due to angular loading can result in higher mechanical and thermal efficiency. Because of the inherently variable compression ratio, the FPLG can run on variety of fuels, making it an appealing platform for energy production in hybrid-electric vehicles and portable power applications [3–7].

A crank-slider mechanism controls piston motion in a traditional ICE, resulting in fixed piston endpoints. The mechanical flywheel's rotational inertia smooths out cyclic irregularities in combustion, which aids to overcome engine misfires and fluctuations in combustion energy. In a free-piston ICE, piston action is unrestricted and is not aided by mechanical flywheel. In the face of significantly variable combustion energies, notably misfires, a totally unassisted FPE is unstable, even though free-piston architecture provides for multiple-fuel capability [8, 9].

The FPLG produces electrical power by transforming the mechanical power of the FPE through electromagnetic induction (Faraday's law). Increasing or lowering the output load current is comparable to changing the "braking force" on the piston because electrical energy output correlates to mechanical energy drain from the FPLG. The flow of output electrical power is controlled by an inverter output stage made up of power transistors. This study aims to see if regulating the output load current using pulse-width-modulation (PWM) switching of the inverter stage can help piston motion [2, 10].

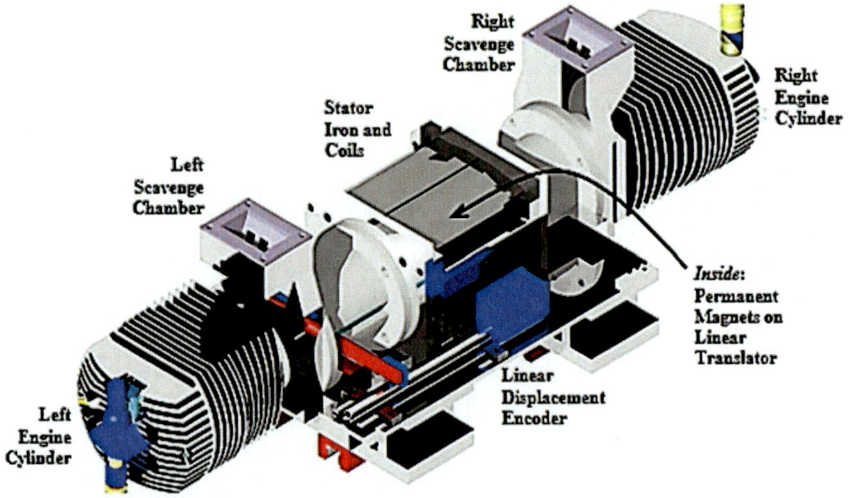

Fig. 1 A free-piston linear electric generator [1]

Table 1 Duration before stoppage for different midpoint CF values and different % COV

	CF = 0.87	CF = 0.89	CF = 0.91	CF = 0.93
10% COV	0.15 s	0.34 s	0.86 s	2.20 s
6% COV	0.15 s	0.33 s	2.14 s	5.89 s

2 System Stability with Various Coefficient of Variation (COV)

Using a simulation model of the FPLG on *Matlab-Simulink*, it has been observed in previous work [10] that consecutively-low combustion factor (CF) values can result in combustion misfire and eventually system stoppage. It is due to low combustion energies and with nominal electrical load current drawn from the system. To overcome this issue, there are two possibilities: either increase combustion energies or decrease electrical load, when the system is starting to produce low combustion energies. In this section, system is analysed for stable operation by variation of combustion energies. This is achieved by changing CF for both 10 and 6% coefficient of variation (COV) to study system behaviour in both cases, as shown in Table 1.

The table shows that for CF of 0.87 and 0.89, the system is dying out almost at the same time of 0.15 s and 0.34 s for both values of COV, respectively. This is because both CF values are still so small and not enough to sustain operation or do not have any impact on sustainability regardless of COV. With lower COV, i.e., at 6%, operation is sustained for longer duration as compared to 10% COV, which is higher than 6% COV. This is because, the lower the COV, the lower is the minimum value of CF. For example, with CF of 0.91, the lowest value of CF is 0.8554 at 6% COV, which is higher than the minimum value of CF at 10% COV, which is 0.819. Thus, the system can sustain for longer with 6% COV.

For CF of 0.91, the system is dying out at 0.86 s for 10% COV and at 2.14 s for 6%. For CF of 0.93, the system dies out after 2.15 s and 5.86 s for COV of 10% and 6%, respectively. When going left to right along each row for both COV 10% and 6%, it is also noted that the higher the CF value, the system will sustain for longer duration but eventually stops after two consecutive low combustion energies. For the sake of simplicity, results for one value of CF = 0.91 at both 10% and 6% COV are shown below. Figure 2 shows graphs of CF and displacement vs time for CF = 0.91 at 10% COV and 6% COV.

3 System Stability with Variation of Load

It is observed in the previous section that consecutively-low combustion energies can result in stoppage of the system. In this section, analysis about variation of load is implemented by reducing the load current when the system is just about to die due to consecutively-low combustion energies. This analysis is done for CF of 0.87 and

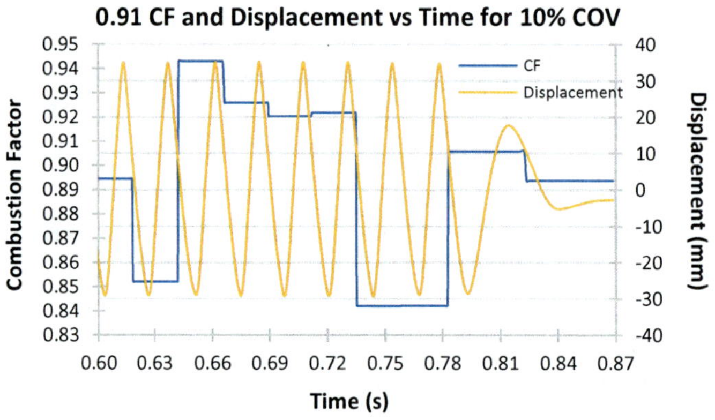

(a) CF=0.91 and 10% COV

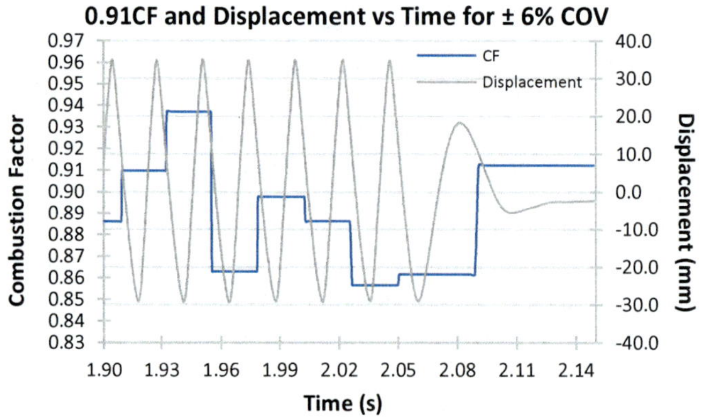

(b) CF=0.91 and 6% COV

Fig. 2 CF and displacement versus time for CF = 0.91 and COV at 10 and 6%

0.89 as the system was dying out so early at 0.15 s and 0.35 s, respectively for 10% COV as discussed. These lowest values of CF are chosen to show that even with such low combustion energy, the system can be made stable with continuous operation by reducing the load current when there are consecutively low combustion energies.

For the sake of simplicity, results for CF 0.87 are shown below. Figure 3 shows the graphs for the impact of reducing load current on the braking force, velocity, displacement, and pressure with CF of 0.87. Current is reduced at the moment when the system is just about to die. It means it is reduced at 0.12 s, which is the last compression stroke inside BC before dying. Previously when the current was 16.55 A, it was dying out just after that time. When current is reduced to 15.30 A as shown in Fig. 3a, it results in reducing of braking force at the same time under equation $F_{elect\_braking}(t) = \frac{K_{\phi}^{2}(x).v(t)}{R}$ as shown in Fig. 3c.

Thus, continuous sustained operation without interruption that was due to a bit higher current is achieved (Fig. 3d). By comparing the results in Fig. 3a and 3d, it can be verified that the system was dying out at around 0.18 s. It can be seen in both figures that due to the reduction of current, the number of cycles from TDC to BDC and back to TDC also increased for the same interval of time. This is because of increased velocity causing more displacement endpoints towards CC side at the

Fig. 3 System Analysis to see the impact of reduced load current on the braking force, velocity, CC and BC Displacement and its endpoints and CC and BC Pressure for values of 0.87

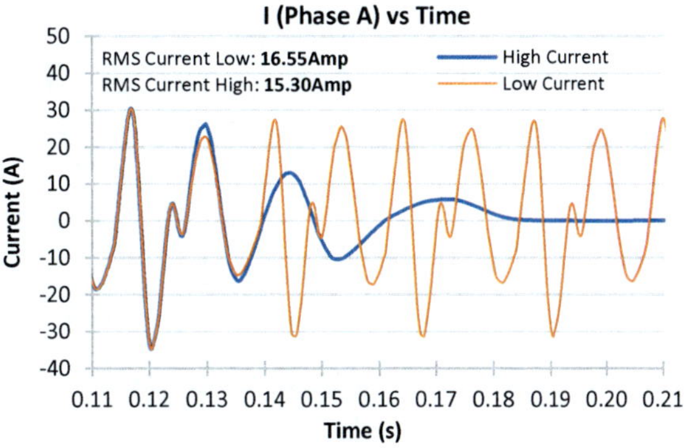

(a) Current vs time

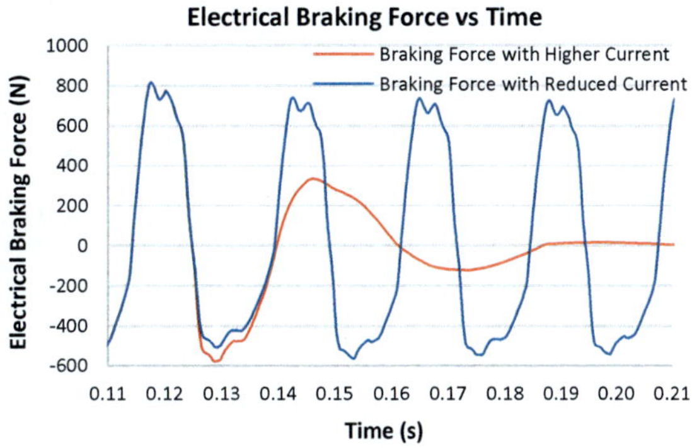

(b) Braking force vs time

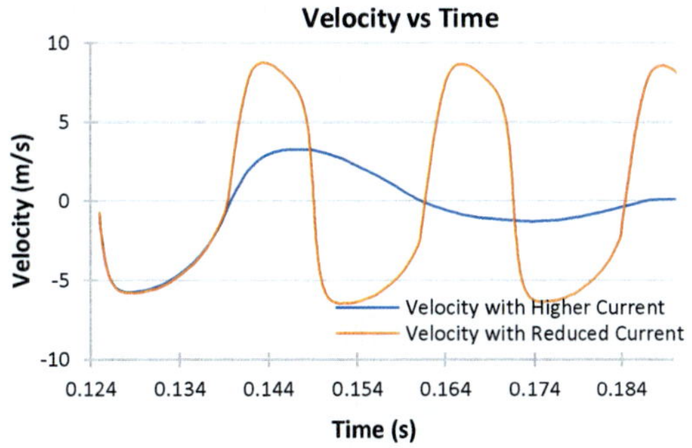

(c) Velocity vs time

Fig. 3 (continued)

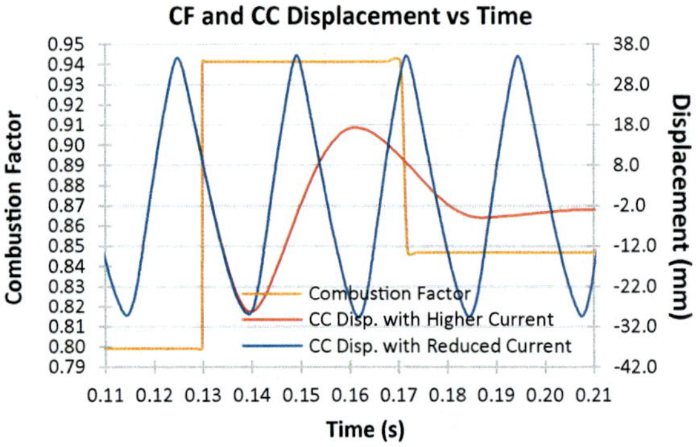

(d) CF & CC disp. vs time

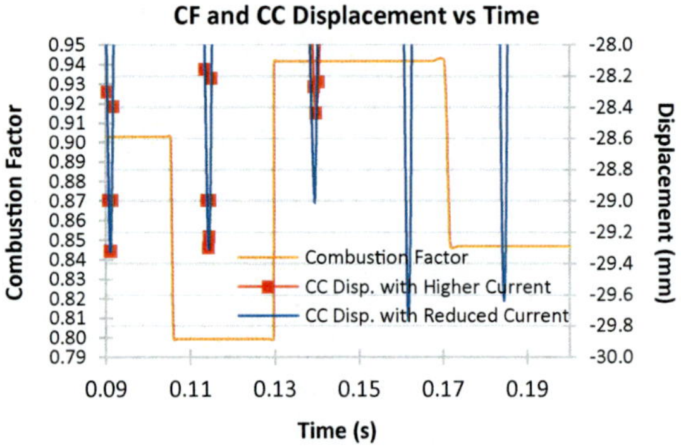

(e) CF & CC disp. endpoints vs time

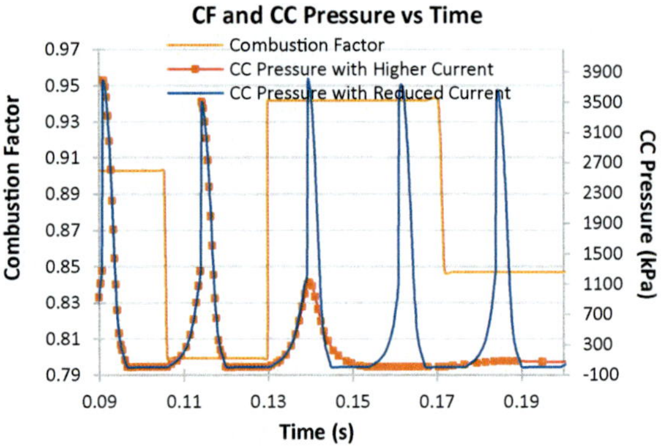

(f) CF & CC pressure vs time

Fig. 3 (continued)

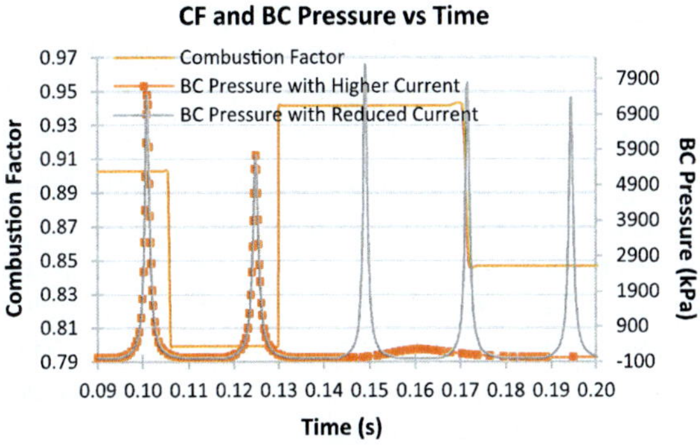

(g) CF & BD Pressure vs time

same time when current is reduced at 0.12 s (Fig. 3e). Thus, healthy combustion is achieved for sustained operation due to more combustion pressure inside. This healthy combustion also results in more displacement endpoints towards BC side and hence more compression pressure (Fig. 3f, g). It helps to have more healthy combustion during return stroke and hence continuous operation is achieved.

4 Control of Load Current Based on Piston Velocity

The research aims to see the viability of steady state operation via varying the load current and braking force. For this, until now, various analysis has been done in different angles to see the impact of varying load current on system stability. This includes analyzing the impact of combustion force and braking force variation on displacement endpoints and piston trajectory. This has been done for single stroke and complete cycle, for no load and full load operation. Similarly, the system is observed with various percentage coefficients of variations (%COV)/combustion variations and load current variations.

In all these parametric variations and load optimization, it is observed that more combustion forces and less current drawn result in more sustained operation with more displacement endpoints. Similarly, less combustion forces and more current drawn result in instability and less displacement endpoints which causes the system stoppage. By observing the time at which the system starts to decay its displacement, the current is switched to lower down at the same time to avoid the system decay and to achieve steady state operation. It is observed that with any value of %COV, if the current is reduced just before the system starts to stop, it can be safe from drying out.

Hence, it was concluded that load current variations (more or less drawn from the system) have a remarkable impact on system stability in displacement endpoints and continuous operation. It means that if the electrical current drain is controlled based on displacement and velocity, steady state continuous operation can be achieved. Based on these observations, a control algorithm has been designed.

The initial control algorithm is implemented on both mathematical and actual electrical subsystems. In mathematical modelling, R is the parameter (equation $F_{elect\_braking}(t) = \frac{K_\phi^2(x).v(t)}{R}$), which is modulated, to vary the load current. While in the existing system, the load current is modulated to ensure the piston achieves the minimum displacement (−29 mm) endpoint to healthy combustion. The algorithm measures the piston velocity once every cycle, at around the displacement of −11 mm and when the velocity is +ve (going from CC to BC). This means new combustion has just occurred, and since combustion energy cannot be directly measured, piston velocity is an indicator of the system energy.

The velocity is compared with a minimum value of velocity, taken at an 8.5 m/s decrease. If the velocity is lower than the minimum velocity, R (for mathematical subsystem) will increase, or duty cycle (for actual electrical subsystem) will decrease. This assists the piston in achieving a minimum displacement of −29 mm so that operation will be sustained. Control on mathematical subsystem model:

R has a nominal value of 3 Ω for the mathematical model. There is a proportional relationship between the value of R and the difference between minimum velocity and measured velocity and is based on how much the measured velocity is lower than minimum velocity, which is 8.5 m/s. The equation is given by:

$$R_T = R_n + K(\frac{8.5m}{s} - v) \tag{1}$$

Value of K is taken as 2, and hence Eq. (1) becomes,

$$R_T = 3 + 2\left(\frac{8.5m}{s} - v\right) \tag{2}$$

Different values of v have been taken for the sake of analysis to see the corresponding value of R_T. The graph is drawn to see the impact of reduced velocity than minimum velocity (8.5 m/s). As the value of R_T is directly proportional to the ΔR, hence R_T will increase in case velocity tries to reduce than 8.5 m/s to reduce the braking force (equation $F_{elect\_braking}(t) = \frac{K_\phi^2(x).v(t)}{R}$), to achieve steady-state operation. This can be seen from Table 2 and Fig. 4 below:

It can be seen from the graph and table that if velocity tries to reduce than 8.5 m/s which is the minimum critical velocity to keep sustained operation, it will result in the reduction of electrical braking force by increasing the resistance R_T, which is the same resistance given in equation for braking force to have a continuous operation. The control algorithm is designed to switch to a higher value of resistance only if it detects the velocity value lower than 8.5 m/s. For all other values of velocities more than 8.5 m/s, it will give a nominal value of resistance 3 ohms. The control algorithm is shown in Fig. 5.

Table 2 Midpoint CF value with unique CF random pattern

v (m/s)	$R_T(ohm)$
8.5	3.0
8.4	3.2
8.3	3.4
8.2	3.6
8.1	3.8
8.0	4.0
7.9	4.2

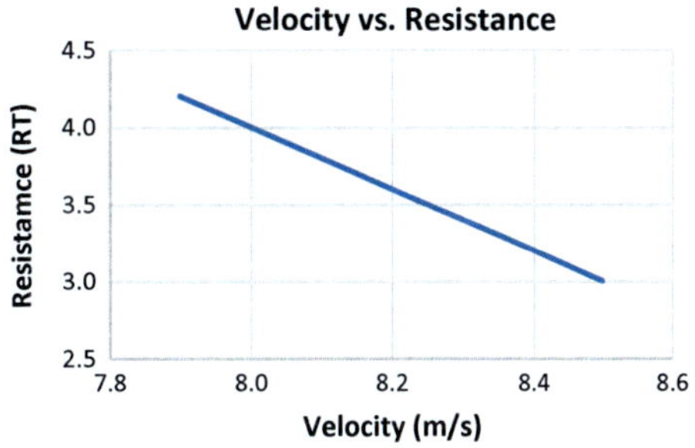

Fig. 4 Relationship between velocity and resistance

Graphs have been drawn to see the automatic resistance variation (current) to achieve the desired motion and velocity profile. This automatic variation is based on the fact that whenever velocity tries to be below the minimum value of 8.5 m/s or displacement tries of being less than −29 mm on CC side is detected (because of low combustion energies), it results in a change in resistance (current) for the sustained

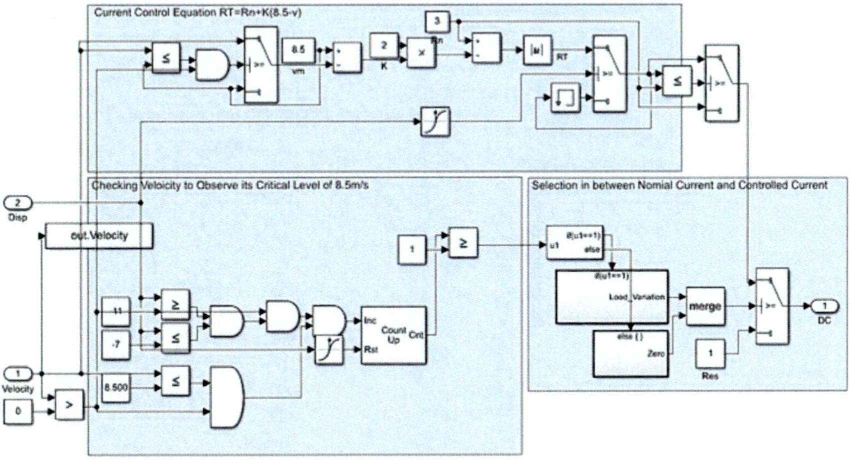

Fig. 5 Initial control algorithm for the mathematical model

operation to reduce the braking force, by Eq. (2). Figure 6 shows resistance variation in accordance with velocity and displacement to adjust the suitable current value for sustained operation. CF of 89 with 10% COV is taken. It is observed that when CF is low, the system tries to die out. To overcome these low combustion energies and avoid the system stopping, resistance is increased to reduce the current drawn.

Fig. 6 Resistance with Combustion Factor (CF), velocity, displacement, current versus time

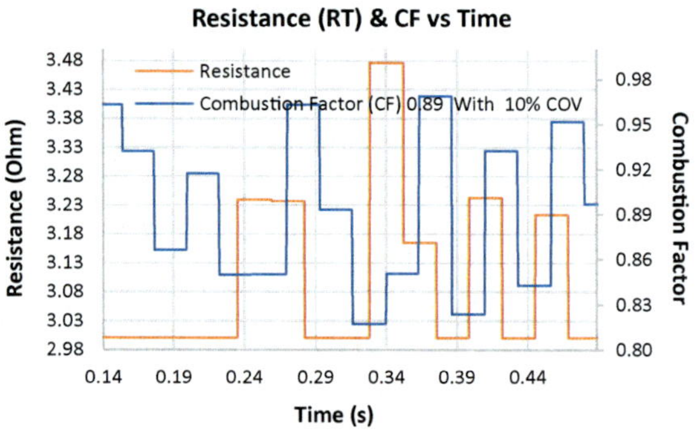

(a) Resistance & Combustion Factor (CF 0.89) vs time

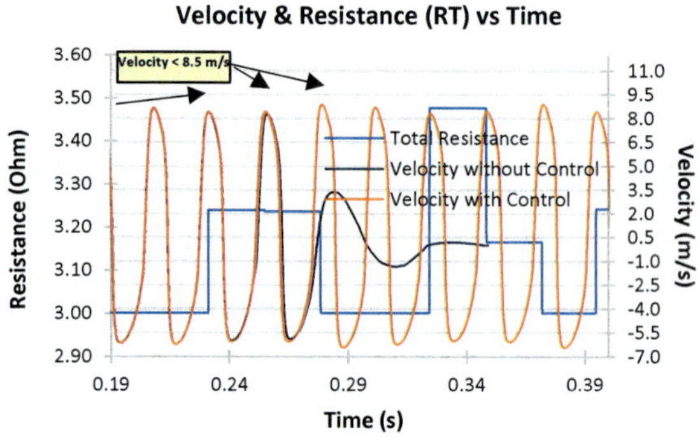

(b) Resistance & velocity vs time

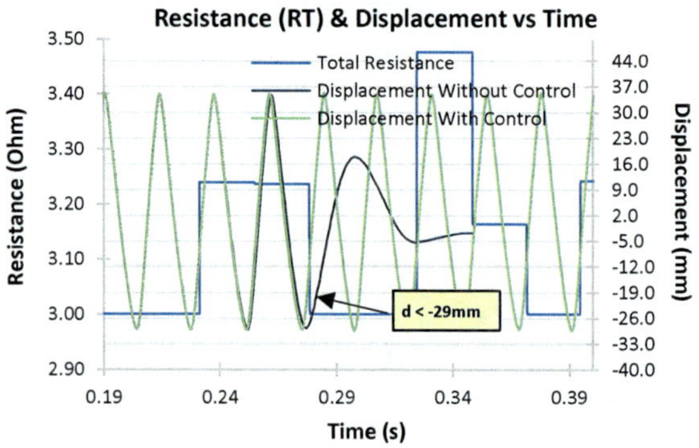

(c) Resistance & displacement vs time

Fig. 6 (continued)

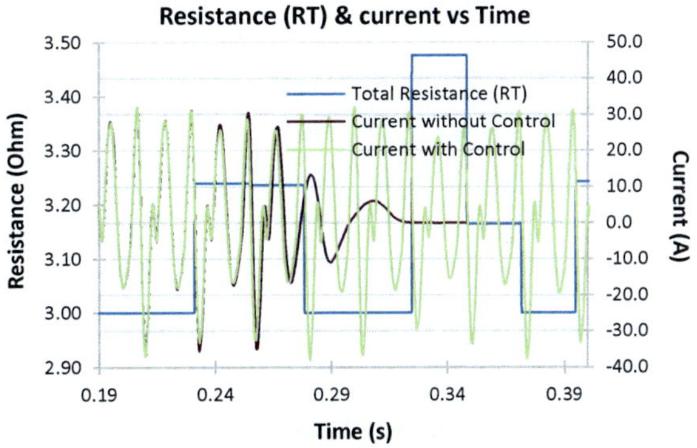

(d) Resistance & current vs time

Figure 6a shows that when low combustion energies are detected, they will increase the resistance (reduce the current) to avoid the velocity dropping below the critical level or displacement below the minimum level to achieve the desired compression ratio for healthy combustion. Due to low combustion energies (from 0.231 s to 0.79 s), velocity becomes lower than 8.5 m/s (Fig. 6b) and displacement also becomes lower than −29 mm (Fig. 6c). At the same time, resistance increase (from 0.231 s to 0.79 s) to reduce the current during the same time period (Fig. 6d). This results in sustained operation without interruption.

5 Conclusion

This research aims to investigate viability of varying the load current and braking force of the FPLG to achieve steady-state operation. Various analysis has been done to assess the impact of varying load current on system stability. By observing the time at which the system starts to decay in displacement, the load current is lowered to avoid system decay and achieve continuous operation. It is observed that with any value of %COV, if the current is reduced just before the system starts to decay, it can be saved from dying out with any value of %COV. Initial control algorithm is implemented on the mathematical model for the linear generator subsystem. The aim is to assist the piston to achieve the minimum displacement of −29 mm to sustain operation. Whenever low combustion energies are detected, the resistance in increased to reduce the current to avoid the displacement below the minimum level. Due to low combustion energies, velocity becomes lower than 8.5 m/s and displacement also becomes lower than −29 mm. When the resistance is increased to reduce the current drawn and thus reduce the electrical braking force, this will result in sustained operation.

Acknowledgements Funding for this research is provided by Petroleum Research Fund (PRF) research grant (0153AB-A34) and the Electrical and Electronics Engineering Department at UTP.

References

1. Zulkifli SA, Karsit MN, Aziz ARA (2008) Rectangular current commutation and open-loop control for starting of a free-piston linear engine-generator. PECon 2008 - 2008 IEEE 2nd International Power and Energy Conference, pp 1086–1091. https://doi.org/10.1109/PECON.2008.4762637
2. Liu D, Xu Z, Liu L, Zhang L, Zhou H (2018) Modeling and control of a portable two-stroke free-piston engine generator. In: Proceedings of 2018 IEEE international conference on mechatronics and automation, ICMA 2018. IEEE, pp 2316–2321. https://doi.org/10.1109/ICMA.2018.8484554
3. Mikalsen R, Roskilly AP (2008) The design and simulation of a two-stroke free-piston compression ignition engine for electrical power generation. Appl Therm Eng 28:589–600. https://doi.org/10.1016/j.applthermaleng.2007.04.009
4. Zulkifli SA, Mohd S, Saad N, Rashid A, Aziz A (2014) Impact of motor size & efficiency on acceleration, fuel consumption & emissions of split- axle through- the- road parallel hybrid electric vehicle. Appl Mech Mater 663:498–503. https://doi.org/10.4028/www.scientific.net/AMM.663.498
5. Moriya K, Goto S, Akita T, Kosaka H, Hotta Y, Nakakita K (2016) Development of free piston engine linear generator system Part3 -novel control method of linear generator for to improve efficiency and stability. SAE Technical Papers 2016-April. https://doi.org/10.4271/2016-01-0685
6. Bashir A, Zulkifli SA, Abidin EZZ, Aziz ARA (2020) Control of Piston Trajectory in a free-piston linear electric generator by load current modulation, pp 465–470. https://doi.org/10.1109/scored50371.2020.9250983
7. Hung NB, Lim O (2016) A review of free-piston linear engines. Applied Energy 178. Elsevier Ltd, pp 78–97. https://doi.org/10.1016/j.apenergy.2016.06.038
8. Petrichenko D, Tatarnikov A, Papkin I (2015) Approach to electromagnetic control of the extreme positions of a piston in a free piston generator. Mod Appl Sci 9:119–128. https://doi.org/10.5539/mas.v9n1p119
9. Zhang C, Chen F, Li L, Xu Z, Liu L, Yang G, Lian H, Tian Y (2018) A free-piston linear generator control strategy for improving output power. Energies 11:1–21. https://doi.org/10.3390/en11010135
10. Bashir A, Zulkifli SA, Aziz ARA, Abidin EZZ (2021) Impact of combustion variance on sustainability of free-piston linear generator during steady-state generation. Energies 14:1–21. https://doi.org/10.3390/en14144081

GBI-Based Wireless Home Automation System Using IoT

Anas Javaid, Anis Fariza Md. Pazil, Chong Hock Siong, and Nurulazlina Ramli

Abstract This paper proposed the design and implementation of Green Building Index (GBI)-based of home automation for monitoring and controlling with safety features integrated by the adoption of Internet of Things. The GBI criteria encompasses the energy efficiency, indoor environmental quality, water efficiency and innovation were implemented for this project focusing on lighting, heating, and ventilation of indoor and outdoor applications. The integration of smart plant irrigation system with rainwater harvesting system is also proposed for plants watering automation. To complement the GBI-based innovation criteria, the add-on features of smart security system such as facial recognition, movement and flame detection using various sensors were also incorporated into the design. In the end, all of the subsystems are connected into the cloud server for electrical power consumption monitoring performed by the smart energy meter. Finally, a working prototype is developed to address the proposed solutions and has successfully accomplished all tested conditions.

Keywords Green building index · Home automation · Internet of things · Smart security

1 Introduction

The Green Building Index (GBI) is Malaysia's recognized green rating system to promote sustainability in the built environment and raise awareness about environmental issues and responsibility to the future generations [1]. The six GBI assessment for building provides an opportunity for project developers and can be used by homeowners to improve green design and construction to their own properties. When GBI

A. Javaid (✉) · A. F. Md. Pazil · C. H. Siong · N. Ramli
Centre of Advanced Electrical & Electronic System, Faculty of Engineering, Built Environment, and Information Technology, SEGi University, 47810, Petaling Jaya, Selangor, Malaysia
e-mail: anasjavaid@icloud.com

Z. Zakaria and S. S. Emamian (eds.), *Recent Advances in Electrical and Electronic Engineering and Computer Science*, Lecture Notes in Electrical Engineering 865, https://doi.org/10.1007/978-981-16-9781-4_12

features is integrated with Internet of Things (IoT) for the smart home automation system, the end result will be more remarkable as it incorporates sustainability weightage with the giant network connecting human and appliances/devices.

An IoT platform is a technology that allow sharing, control, and automate directly any connected IoT devices. The main concept is to attach various devices to a specific cloud with high-level of security, pliant options of connection, with a wide and fast data processing ability. Today, the increasing demand of automated systems especially for home leads to abundance research and studies being conducted in this area of interest.

The studies in [2] presented the research work of smart home automation and security using the Arduino modules but without surveillance system unlike the work presented in [3] which includes in the surveillance monitoring in their system using the Raspberry Pi board. Nevertheless, the design in [3] might get a bit complicated as it is developed using Linux, Python and Apache Server which needs basic programming skills in the above language. In addition, both systems do not have any considerations at all on GBI-based implementation. Therefore, this project mainly aims to enhance the current wireless home automation system by integrating IoT features with GBI-based criteria to promote green building with technology. This is to encourage the efficient use of natural resources particularly in energy and water while at the same time being supported by robust security features.

2 Methodology

Figure 1 shows the basic block diagram of the proposed methodology made up of 5 main subsystems categories implemented into this project while having a flexibility for the system to add more modular subsystem if needed. The indoor and outdoor lighting and smart system of fire, camera and movement detector are powered by the solar for the system security sustainability in case of power outage. The employability of the dependable wireless technology which connects multiple modules to the home automation system's server; provides greater flexibility the design solution. The goal of this proposed methodology is to develop and implement a cost-effective design of smart home automation system while sustaining the environmentally friendly ambience driven by GBI guidelines.

2.1 System Design

2.1.1 Arduino D1 Wemos

The Wemos D1 development board is based on the popular chip that revolutionized Wi-Fi in embedded systems of serial ESP8266. This simple module allows for system prototyping for the IoT in a less complicated manner as does not require

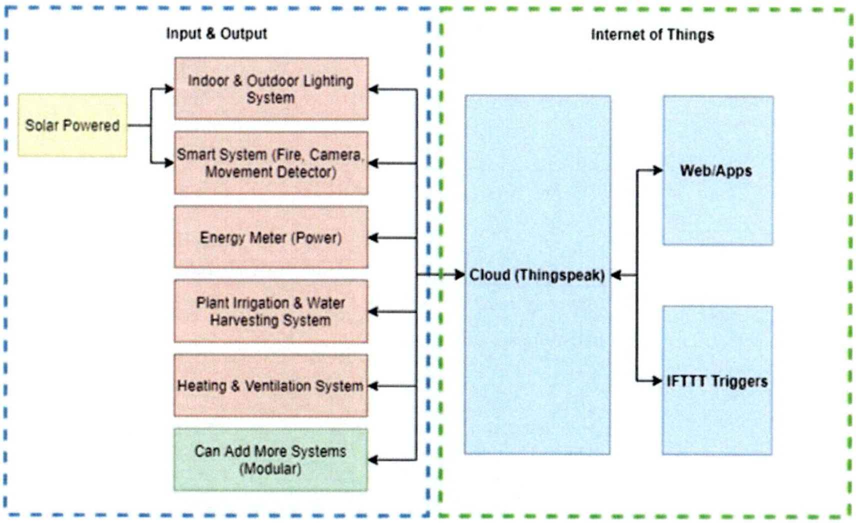

Fig. 1 Basic block diagram of the proposed methodology

other accessories to have an internet connection such as the official Arduino ethernet shield or Wi-Fi shield accessory. In addition, Wemos D1 can be programmed using the Arduino IDE in C and C++ languages which provides an advantage for programming flexibility.

2.1.2 Sensors and Camera Interfacing

Various of sensors are used in this project such as DH11 to measure the humidity and the temperature, flame sensor of LM393 to detect fire and Passive Infrared (PIR) sensor to detect motion of 10 m range. In addition, a soil moisture sensor which is an analog sensor is also used that shows values in voltage by which can be converted into percentage for indicator reference. The Light Dependent Resistor (LDR) sensor used in this project is to calculate the intensity of the light while the water level sensor detects the level of the water touching the surface of the sensor and gives feedback accordingly. To measure the power for the energy system, a voltage sensor (ZMPT101B) and current sensor (ACS712 5A) is integrated to the system. Finally, the camera used in this project is ESP32-CAM, which is based on the ESP32 board that allows direct camera module integration and at the same time used for motion detection and facial recognition.

Table 1 The input and output of GBI-based IoT System Design

GBI criteria	Subsystems	Input	Output
Energy efficiency	Power energy meter	Voltage & current	Voltage & current graph visuals on IoT Cloud
	Lighting system	Smart devices & LDR	Graph visuals, ON/OFF control using mobile, automatic garden lights
	Solar power system	Sunlight	Regulated DC Supply
Water efficiency	Plant irrigation system	Soil moisture sensor	Water pump, moisture graph visuals & data on IoT Cloud
	Water level harvesting	SL067 sensor	Water level visuals on IoT Cloud
Indoor environment quality	Heating & ventilation system	DHT sensor	Auto fan & Exhaust, T & H graphs on IoT Cloud
Innovation	Security & fire alarm system	LM393, Camera, PIR	Buzzer, IoT data Cloud, live-feed & door control

2.1.3 Architecture of the Proposed GBI-Based IoT System Design

Table 1 shows the input and the output of the proposed GBI-based with the related sensors, devices, and equipment for the implementation of this project.

The block diagram in Fig. 2 below shows several blocks supported by the subsystem indicating the connectivity of all sensors, devices and equipment implemented for this project. From the figure shown, all data from the sensors are being collected by the Thingspeak cloud which served as an IoT cloud platform. The data will be further analysed, monitored, and can be used for applications such as '*If This Then That*' (IFTTT), MATLAB Analysis or Webhooks. In this project, IFTT is used as the platform application as it provides simplicity in design implementation.

3 Implementation

3.1 Hardware Implementation

As shown in Fig. 3, the major components of hardware parts used in the subsystems are Arduino microcontrollers, camera, and sensors with relays. Many sensors are interfaced with Arduino through C programming language and the output can be viewed on the cloud server, webpage, and mobile platform.

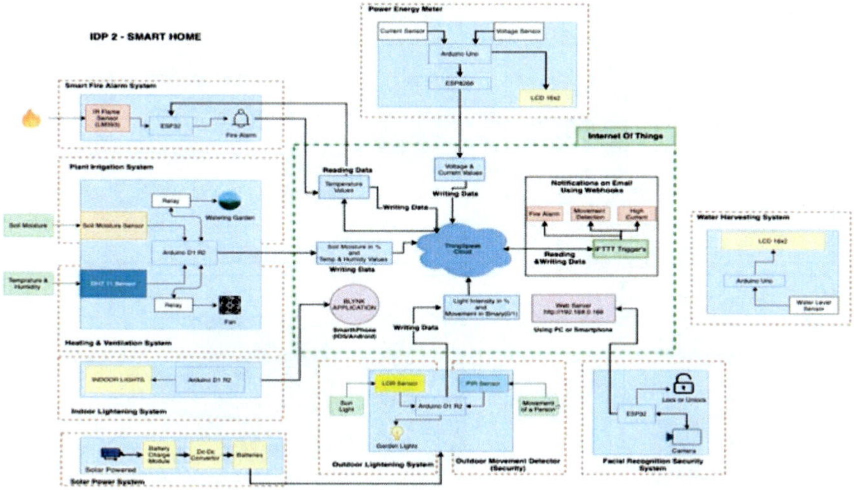

Fig. 2 Block diagram of the integrated architecture for proposed design

3.2 Software Implementation

In this project, Thingspeak online platform is used as it is one of the easiest platforms to set up as compared to other online platforms especially for data configuration. In addition, it also provides a high quality of support. This online platform can visualize data in real time where results can be obtained instantaneously. This means that the data processing can be performed on real time-based with visualization feature for the interaction ability with apps and plugins. It is also low cost as there are options for a free user account. Another software integrated into the design is IFTTT which allows the user to use all data being collected by ThingSpeak and also Blynk app for subsystem control application [4, 6].

4 Result

As shown in Fig. 4, a simplified house prototype has been developed to demonstrate the functionality and the output of each subsystem. The results displayed for all of the subsystems shows that the system has been integrated and developed successfully that met the criteria and objectives of the proposed design.

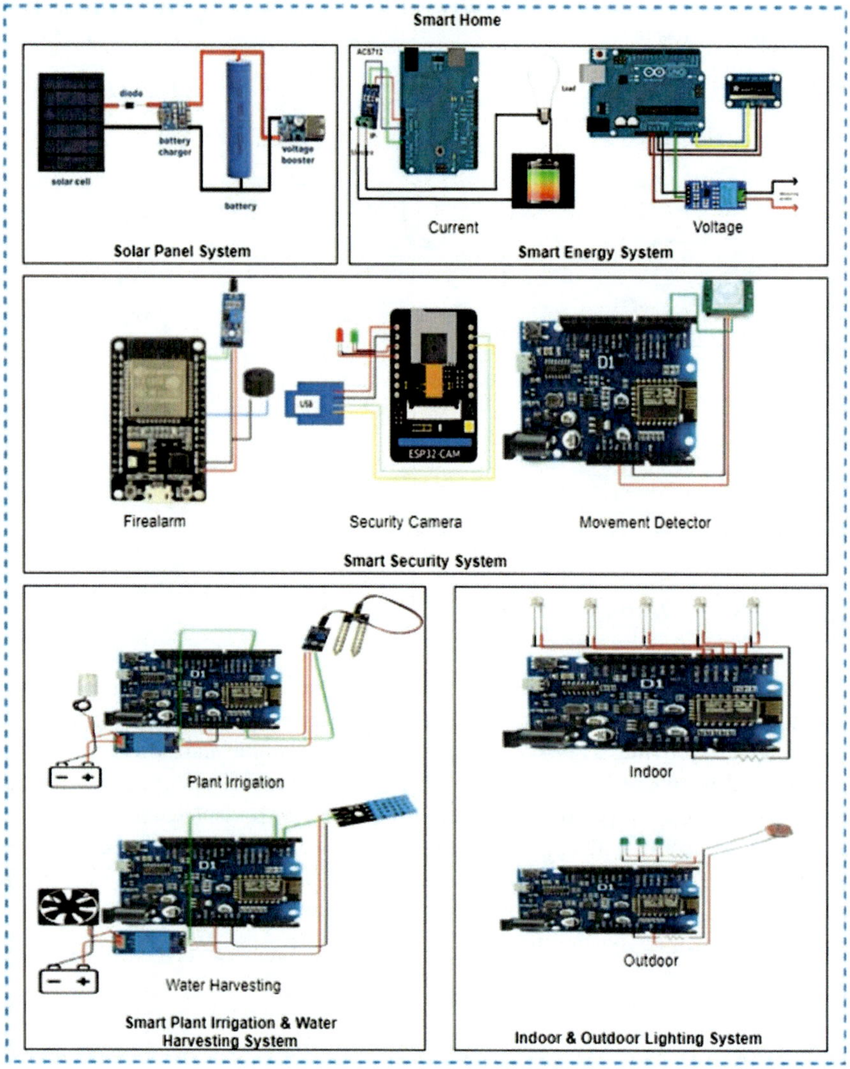

Fig. 3 Hardware implementation

Meanwhile, Fig. 5 shows the display results extracted from the Thingspeak cloud. These panels or dashboard shown will help the user to have a full control of their smart home by monitoring each subsystems functionality from any choice of smart device.

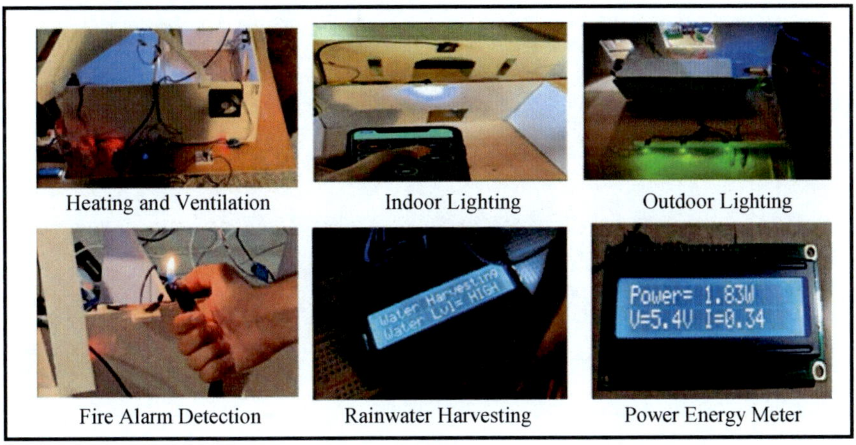

Fig. 4 Subsystem hardware prototype display results

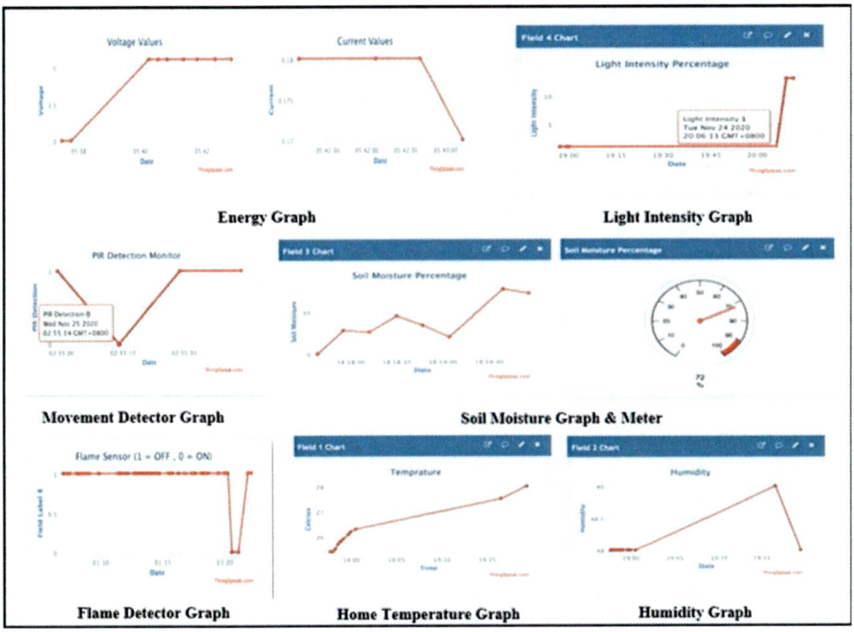

Fig. 5 Subsystems visual display for result monitoring from the online platform

4.1 Security System

In Fig. 6, the results from security camera are shown whereby the top left picture shows intruder alert when the person is not registered to the system while the top right

Fig. 6 Results shown from the security camera

picture shows the person is recognized after being registered. Meanwhile, the house model of picture below showing the red light indicating the intruder is prohibited from entering the house and a green light granting access to the person.

5 Comparison with Existing System

Table 2 shows the comparison of the proposed system with the existing system based on the wireless home automation system application. To the authors' knowledge, no prior studies have been performed so far to integrate the IoT element with the GBI-based features for the wireless home automation system.

Table 2 Comparison between proposed and existing systems

References	Parameters	Existing systems	Proposed system
Gupta and Chhabra [5]	Green technology	Less or none	GBI-based
Santoso and Vun [7]	Circuit integration	Complex	Simplified
Jabbar et al. [8]	User Friendly	Less or none	Highly interactive
Pătru et al. [9]	Modular	None	Upgradable by subsystem(s) addition

6 Conclusion

The GBI criteria meeting the requirements of energy efficiency, indoor environmental quality, water efficiency and innovation has been fully developed in this home automation system with IoT adoption. The application of monitoring and controlling with safety features incorporated into the design by Arduino D1 Wemos and the online platform of ThingSpeak and Blynk app shows an excellent agreement with the aim of the project. This design provides moderate and less expensive way of sensing, monitoring, and controlling system in the most efficient way while considering the sustainability of natural resources of the world. In future, it is expected that more GBI-based criteria such as wastewater recycling management can be incorporated into the design for greater benefits to the ecosystem. While for future design, it can be proposed that the controlling of the systems to be done via voice recognition to enhance the security features of the design.

References

1. Harun MF, Samah AA, Majid HA et al (2017) Optimization of green building design to achieve green building index (GBI) using genetic algorithm (GA). In: 6th ICT International student project conference: elevating community through ICT, ICT-ISPC 2017, 2017-Janua, pp 1–4. https://doi.org/10.1109/ICT-ISPC.2017.8075310
2. Wadhwani S, Singh U, Singh P et al (2018) Smart home automation and security system using Arduino and IOT. Int Res J Eng Technol (IRJET) 5(2):1357–1359
3. Quadri SAI, Sathish P (2017) IoT based home automation and surveillance system. In: Proceedings of the 2017 International conference on intelligent computing and control systems, ICICCS 2017, 2018-Janua, pp 861–866. https://doi.org/10.1109/ICCONS.2017.8250586
4. Al-Ali R, Al-Rousan M (2004) Java-based home automation system, || IEEE Trans. Consum Electron 50(2):498–504
5. Gupta P, Chhabra J (2016) IoT based Smart Home design using power and security management. In: 2016 International conference on innovation and challenges in cyber security (ICICCS-INBUSH). IEEE
6. Sokullu R, Akkaş MA, Demir E (2020) IoT supported smart home for the elderly. Internet of Things
7. Santoso FK, Vun NC (2015) Securing IoT for smart home system. In 2015 international symposium on consumer electronics (ISCE). IEEE
8. Jabbar WA, Alsibai MH, Amran NSS et al (2018) Design and implementation of IoT-based automation system for smart home. In: 2018 International symposium on networks, computers and communications (ISNCC)
9. Pătru I, Carabaş M, Bărbulescu M at al (2016) Smart home IoT system 2016 15th RoEduNet conference: networking in education and research

Embedded Machine Learning on a Programmable Neuromorphic Platform

Indar Sugiarto, Agustinus Bimo Gumelar, and Astri Yogatama

Abstract This paper presents an implementation of k nearest neighbor (k-NN) algorithm on SpiNNaker. SpiNNaker is a programmable neuromorphic platform well-known for its very low power energy consumption which is suitable to be used as an embedded system. By utilizing the SpiNNaker communication protocols, we are able to efficiently distribute jobs across SpiNNaker cores for performing the k-NN algorithm. From the experiments, we observed that the k-NN program runs smoothly and performs basic classification task on Irish dataset correctly. From the investigation, we found that the k-NN program reports not only the accuracy and the best k-value to get that accuracy, but also the reason why that k-value should be chosen for the corresponding classification task. We also found that increasing the potential number for k-values, the k-NN program was able to find better accuracy. For example, the accuracy of 91.67% was achieved when we increased the range of k between 30 to 60.

1 Introduction

Machine learning (ML) is a technique for exploring and exploiting data that can be used to improve performance of a system or an application. It is considered as the most progressive aspect of artificial intelligence, since many applications nowadays use machine learning algorithms or at least interact with a cloud service provider that provides machine learning capabilities [1].

I. Sugiarto (✉)
Electrical Engineering, Petra Christian University, Surabaya, Indonesia
e-mail: indi@petra.ac.id

A. B. Gumelar
Computer Science, Narotama University, Surabaya, Indonesia
e-mail: bimogumelar@narotama.ac.id

A. Yogatama
Communication Science, Petra Christian University, Surabaya, Indonesia
e-mail: astri@petra.ac.id

Z. Zakaria and S. S. Emamian (eds.), *Recent Advances in Electrical and Electronic Engineering and Computer Science*, Lecture Notes in Electrical Engineering 865, https://doi.org/10.1007/978-981-16-9781-4_13

Machine learning algorithms can also be used in various applications the embedded world. The term edge computing now becomes increasingly popular especially in the context of industrial revolution 4.0 related topics such smart city, smart transportation, and so on. In this sense, many embedded systems are equipped with hardware modules capable of computing or implementing certain machine learning algorithm [2].

One requirement and also the benefit of an edge computing system is the low power consumption whilst maintaining simple operability due to its direct access to sensors and actuators of the associated cyber physical systems [3].

On the other side of emerging technologies, now exist a new computing paradigm called neuromorphic computing. This paradigm exhibits superiority on low power requirement of an artificial intelligence (AI) based computing [4]. This neuromorphic computing is supported by various neuromorphic devices. Along with their software architecture, these neuromorphic devices comprise the so called neuromorphic platform. The presence of neuromorphic platforms seems to be inline with the advancing trend of edge computing. Together, these technologies may offer hugh benefits in term of inter-operability as well as reliability of an AI-driven ecosystem in the future [5].

This paper presents preliminary results of experiments on implementing AI concept in a programmable neuromorphic platform. In this study, some ML algorithms were implemented and tested on a SpiNNaker platform. SpiNNaker, which stands for Spiking Neural Network Architecture, originally was designed for implementing the third generation of artificial neural network [6]. By implementing ML algorithms on SpiNNaker, we gain not only the low power energy consumption, but also high configurability of ML-based applications, since the SpiNNaker offers many options to optimize the program running on it [7]. In this paper, only the k-NN (k nearest neighbor) algorithm is reported.

This paper is organized as follows. After giving some introductory background, we present our implementation method in Sect. 2. The experiment results and the discussion are presented in Sect. 3. Finally, this paper is closed by conclusion presented in Sect. 4.

2 Methods

2.1 Brief Overview of SpiNNaker

SpiNNaker is originally a neuromorphic platform dedicated for the study of human brain in digital form. It was conceived by the University of Manchester in the UK, where computer scientists work together with computational neuroscientists to create a system capable of simulating millions of digital neurons in real time [8].

SpiNNaker ecosystem consists of various level of neuromorphic computing, from the chip level neuromorphic devices to the system level brain simulator. The lowest level of this system is the SpiNNaker chip. This chip is a multicore system that has 18 ARM968 cores. These cores are packed along with 256 MB dynamic RAM during the chip production. In addition to this low-power SDRAM, each SpiNNaker core is also equipped with two types of static memory, namely 32 KB ITCM (instruction tightly-coupled instruction memory) exclusively for storing program instruction, and 64 KB DTCM (tightly-coupled data memory).

From these chips, higher level SpiNNaker boards are produced. Currently, there are two type of SpiNNaker boards available for research (see Fig. 1): 4-chips SpiNNaker boards called Spin3 and 48-chips SpiNNaker boards called Spin5.

Since each SpiNNaker chip consists of 18 cores, thus in total, Spin3 will have 72 cores, whereas Spin5 will have 864 cores. Unfortunately, not all cores are usable for normal user's program since SpiNNaker system requires at least one core for internal use.

On both SpiNNaker boards, there is a special SpiNNaker chip (usually designated as chip <0,0> in the SpiNNaker datasheet) that has a special function as to provide the communication access to the outside world through a standard TCP/IP connection. In this paper, as in other our previous papers, we use this special chip as the master that coordinates all SpiNNaker chips during algorithm/program deployment.

There are several modes of communication within the SpiNNaker board. Two of the most important ones for user program deployment are so called SpiNNaker

(a) (b)

Fig. 1 Two types of SpiNNaker boards: **a** Spin3 with 4 SpiNNaker chips, and **b** Spin5 with 48 SpiNNaker chips

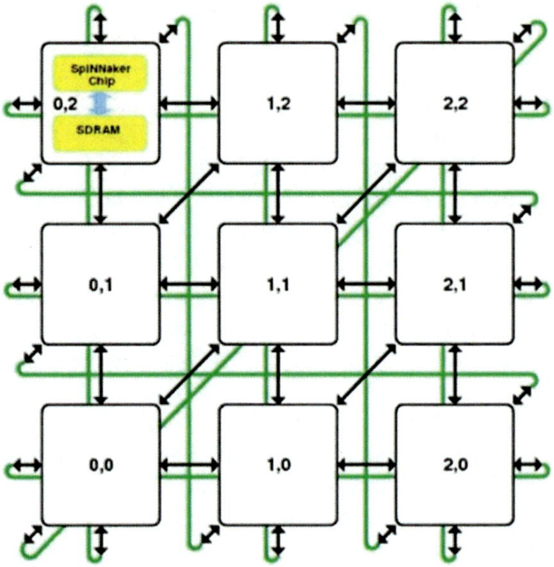

Fig. 2 Each SpiNNaker chip has six bidirectional links that can be connected to other chips. Using these links, 2D triangular mesh topology can be created that makes SpiNNaker an efficient high-throughput many-core system

multicast packet (MC) and SpiNNaker datagram packet (SDP). These two packets are used in this study for different purpose: the MC packets are used for communication and coordination between cores of SpiNNaker, whereas SDP packets are used for delivering data from/to computer host via TCP/IP. The MC packets have higher throughput than SDP packet since they have lower overhead and can optimally make use of the SpiNNaker six channels torus topology as shown in Fig. 2.

In the following subsection, we describe how two machine learning algorithms are implemented on SpiNNaker. In both implementation strategies, we designated one core as a master core. This master core resides in chip <0,0> and has the core virtual number 1; hence, it is often called as core <0,0,1> in the SpiNNaker ecosystem (Fig. 3).

2.2 *K-NN Implementation*

k-NN is a basic ML algorithm that is used commonly in regression or classification tasks. It works pretty simple and can be explained intuitively since it is a non-parametric statistical algorithm that processes k closest training examples.

k-NN on SpiNNaker is implemented by dividing tasks between master core and slave core. The master core is tasked with delegating calculation functions to other cores (referred to as slave cores). In addition, the master core will receive the distinct k-ID of the corresponding slave, and the accuracy value of that slave core. The slave core is in charge of running the k-NN core function and calculating the accuracy

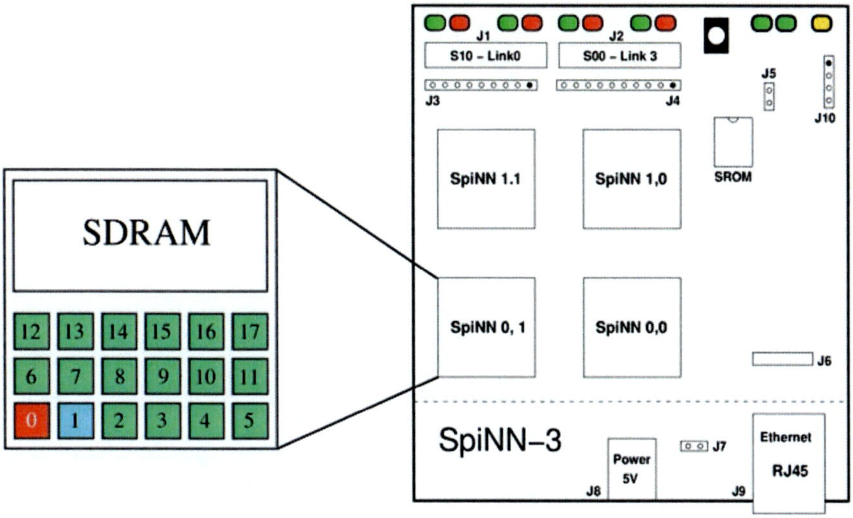

Fig. 3 Each SpiNNaker chip has 18 cores (see Fig. 1 for the real/physical appearance of the Spin3 board). Core-0 is dedicated for SpiNNaker kernel, and core-1 will be used as the master core in our ML implementation strategy. The rest of the cores will be used as worker/slave cores. The master core in chip <0,0> also operates as the master core for the entire system since it has responsibility to communicate with external devices through the ethernet port

value of the given k-value. After computing the accuracy value, that value will be sent to the master core.

The algorithm of the k-NN program on SpiNNaker is designed as follows:

1. Set the routing table to deliver MC packets from the master to slave cores.
2. Set a timer to send a multicast packet from the master core to the slave core. Here the timer is set to run every 1 (one) second and will stop after receiving stop comand from the host PC.
3. Master core sends multicast packet to slave core with parameter k-value.
4. Slave cores listen and receive multicast packets from master.
5. Slave cores perform k-NN calculations based on its predefined region, and get the accuracy value for the k-value obtained.
6. Slave cores sends accuracy and k-value back to the master as a MC packet.
7. Master core receives the accuracy value and k-value from all other slave cores. If the accuracy value received is better than the previous accuracy value, the accuracy value and k-value will be updated.
8. If the timer has stopped (after some iteration or by request), then the process on the slave cores will stop and the master core will send the stored accuracy and k-value to the external device (i.e., host-PC) via SDP.

Figure 4 represents the flowchart of k-NN program running on SpiNNaker. In this study, a standard host-PC is used to send and retrieve data to/from the SpiNNaker

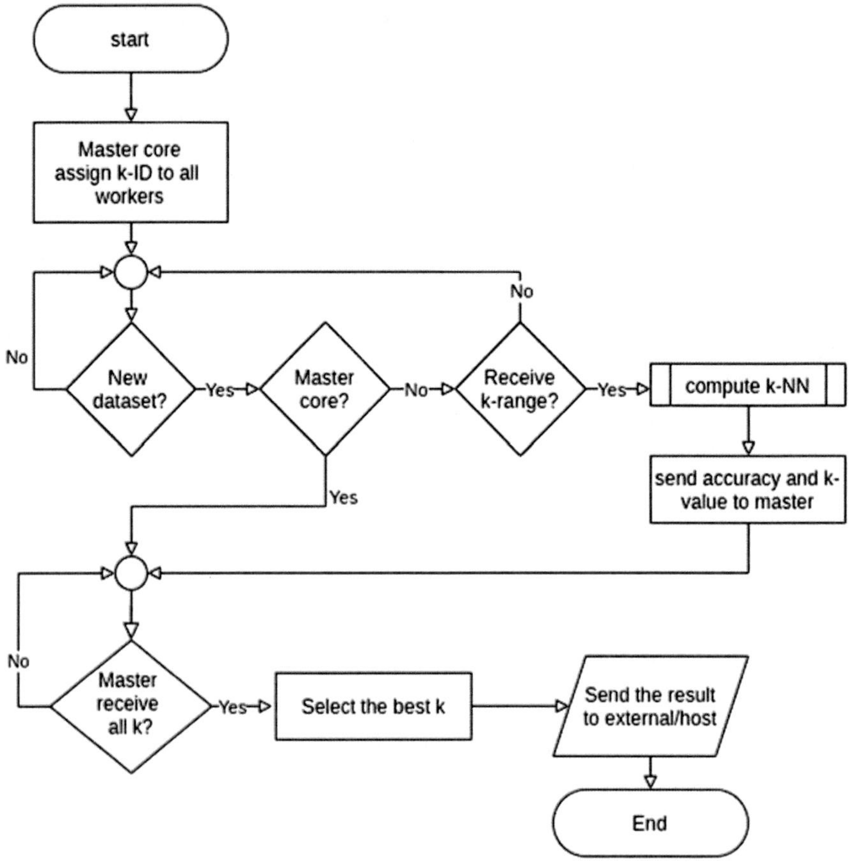

Fig. 4 The main flowchart of k-NN program running on SpiNNaker

board. This host-PC can be replaced by any edge computing hardware in real applications as long as that hardware has a TCP/IP connection to the SpiNNaker board. To receive the results from the SpiNNaker, a Python program is used and it will listen to the datagram socket coming from the SpiNNaker board.

3 Result and Discussion

3.1 K-NN Experiment

For evaluation purpose, we used k-NN for classification of Iris flower using dataset available from the University of California, Irvine. This is a well-known multivariate dataset used by statistician that originated from Ronald Fisher paper in 1936. The dataset consists of 50 species of Iris flower, namely: Iris setosa, Iris virginica and Iris versicolor. In the dataset, four features are used to distinguish the species: the length and width of the sepals and petals, in centimeters.

After deploying the k-NN program onto the SpiNNaker board, we inspected the SpiNNaker to see if everything runs properly. The initial output of the SpiNNaker program shows that the k-NN program was deployed successfully (see Fig. 5).

A Python script has been developed to control how the k-NN program on the SpiNNaker should work. This Python script is also used to notify the k-NN program on the SpiNNaker how many k-values should be used by the k-NN program. This in

```
# ybug - version 3.4.1
SPIN3:0,0,0 > @ knn_spinnaker.ybug
@SPIN3:0,0,0 > iptag 1 set . 17894
@SPIN3:0,0,0 > app_load knn_spinnaker.aplx all 1-17 22
@SPIN3:0,0,0 > sleep 1
@SPIN3:0,0,0 > app_sig all 22 sync0
SPIN3:0,0,0 > sp
SPIN3:0,0,0 > ps
Core State  Application      ID   Running  Started
---- -----  -----------      --   -------  -------
  0  RUN    scamp-3           0   0:24:07   1 Nov 17:33
  1  RUN    knn_spinnaker    22   0:00:07   1 Nov 17:57
  2  RUN    knn_spinnaker    22   0:00:07   1 Nov 17:57
  3  RUN    knn_spinnaker    22   0:00:07   1 Nov 17:57
  4  RUN    knn_spinnaker    22   0:00:07   1 Nov 17:57
  5  RUN    knn_spinnaker    22   0:00:07   1 Nov 17:57
  6  RUN    knn_spinnaker    22   0:00:07   1 Nov 17:57
  7  RUN    knn_spinnaker    22   0:00:07   1 Nov 17:57
  8  RUN    knn_spinnaker    22   0:00:07   1 Nov 17:57
  9  RUN    knn_spinnaker    22   0:00:07   1 Nov 17:57
 10  RUN    knn_spinnaker    22   0:00:07   1 Nov 17:57
 11  RUN    knn_spinnaker    22   0:00:07   1 Nov 17:57
 12  RUN    knn_spinnaker    22   0:00:07   1 Nov 17:57
 13  RUN    knn_spinnaker    22   0:00:07   1 Nov 17:57
 14  RUN    knn_spinnaker    22   0:00:07   1 Nov 17:57
 15  RUN    knn_spinnaker    22   0:00:07   1 Nov 17:57
 16  RUN    knn_spinnaker    22   0:00:07   1 Nov 17:57
 17  RUN    knn_spinnaker    22   0:00:07   1 Nov 17:57
SPIN3:0,0,0 > █
```

Fig. 5 The result of deploying k-NN program onto the SpiNNaker board. Here core-1 operates as the master, and the remaining cores run as the worker (except core-0, which is the kernel of the SpiNNaker system called SCAMP or SpiNNaker Control and Monitor Processor)

turn will determine how many cores will be involved in the k-NN computation. For example, if the number of k was specified as 40, then only 40 SpiNNaker cores will be activated. This mechanism is depicted in Fig. 6.

After feeding the training and testing dataset to the k-NN program on the SpiNNaker, we collected the accuracy values. Table 1 shows the result.

As we can see from Table 1, the accuracy increases as the k-value also increases. However, the increase in accuracy is not monotonically linear and tends to increase step-wisely. The first column in Table 1 represents the range of k values that will be evaluated. This column also represents the total number of SpiNNaker cores involved in the k-NN computation. The second column in Table 1 represents the k-value among the range of active k values that produces the highest accuracy. As we can see in that table, when we varied k values between 6 and 25, the highest accuracy was found at ke equals 6. It means that the ideal k for the given dataset (Irish flower dataset) is 6. However, we also see that when we increased the potential number of k between 30 and 60, we found that the highest accuracy was achieved when the k-value is 29. For the lower range of k-values (between 3 and 5), we see that the optimal number of k tends to the the maximal number of k available in that range.

To understand this behavior further, we look into the distribution of accuracy for the corresponding k-values in the given range. Figure 7 shows how the accuracy was achieved by each k-value in that range.

As we can see on Fig. 7, the highest accuracy was achieved for the k-value number 29. On that figure, we also plot the approximated trending line of the average accuracy across all possible k-values. We found that the knee (or elbow) phenomenon happens

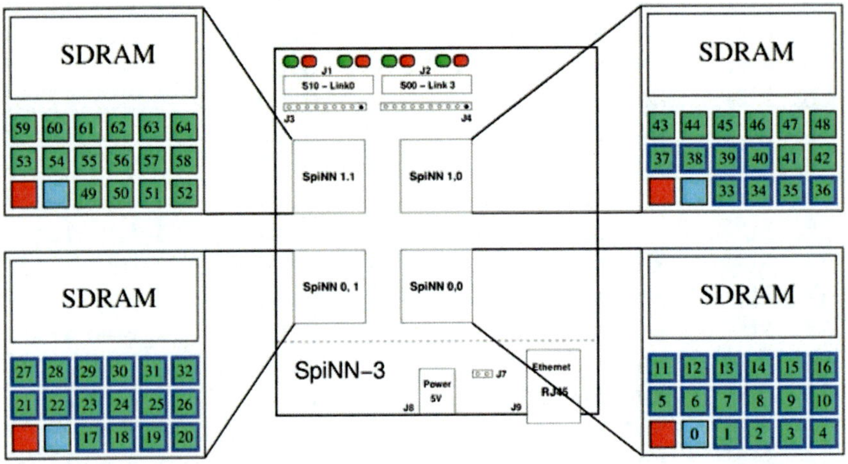

Fig. 6 Example of running k-NN program on SpiNNaker with k = 40. In this case, only 40 cores will be activated as working cores (see the cores with thick blue border above) and the remaining cores will be idle. Core-0 (at chip <0,0>) will act as the master that coordinates all of active working cores

Table 1 The accuracy of k-NN program on SpiNNaker

Max k-range	Optimal k-value	Accuracy (%)
3	3	77.08
4	3	77.08
5	5	85.42
6	6	87.5
7	6	87.5
8	6	87.5
9	6	87.5
10	6	87.5
15	6	87.5
20	6	87.5
25	6	87.5
30	29	91.67
40	29	91.67
50	29	91.67
60	29	91.67

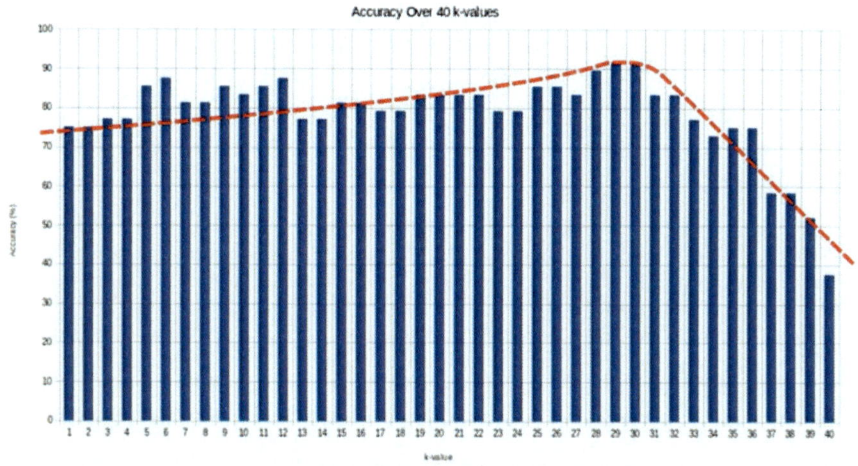

Fig. 7 Achieved accuracy for each possible k-values in the range of 40 k-values. This also corresponds to the distributed k-value across worker cores

at k-29. In ML literature, this corresponds to the elbow method for finding optimal k-value [9].

4 Conclusions

This paper presents a preliminary but novel work on machine learning on SpiNNaker. SpiNNaker is a programmable neuromorphic platform which is basically a very low power many core system. We implemented a basic machine learning algorithm called k nearest neighbor (k-NN) that is commonly used in classification and regression tasks. By utilizing two type of SpiNNaker communication packets, namely MC-packets and SDP, we are able to efficiently distribute jobs across SpiNNaker cores for performing k-NN algorithm. From the experiments, we observed that the k-NN program runs smoothly and performs basic classification task on Irish dataset correctly. From the investigation, we found that the k-NN program reports not only the accuracy and the best k-value to get that accuracy, but also the reason why that k-value should be chosen for the corresponding classification task. We also found that increasing the potential number for k-values, the k-NN program was able to find better accuracy. For example, the accuracy of 91.67% was achieved when we increase the range of k between 30 to 60. For our future work, we will implement other ML algorithms such as SVM, MLP (multilayer perceptron), etc., as well as testing those algorithms on real world application scenarios.

Acknowledgements This research was supported by the Directorate of Higher Education of the Ministry of Research and Technology of the Republic of Indonesia (Ristekdikti) through the grant Penelitian Dasar (PD) no: 002/AMD-SP2H/LT-MULTI-PDPK/LPPM-UKP/2021.

References

1. Neloy AA, Alam S, Bindu RA, Moni NJ (2019) Machine learning based health prediction system using IBM Cloud as PaaS. In: 2019 3rd international conference on trends in electronics and informatics (ICOEI), pp 444–450
2. Branco S, Ferreira AG, Cabral J (2019) Machine learning in resource-scarce embedded systems, FPGAs, and end-devices: a survey. Electronics 8(11):1289
3. Shi W, Dustdar S (2016) The promise of edge computing. Computer 49(5):78–81
4. Marković D, Mizrahi A, Querlioz D, Grollier J (2020) Physics for neuromorphic computing. Nat Rev Phys 2(9):499–510
5. Mead C (2020) How we created neuromorphic engineering. Nat Electron 3(7):434–435
6. Furber S (2016) Large-scale neuromorphic computing systems. J Neural Eng 13(5):051001
7. Sugiarto I, Furber S (2021) Fine-grained or coarse-grained? Strategies for implement-ing parallel genetic algorithms in a programmable neuromorphic platform. TELKOMNIKA Telecommun Comput Electron Control 19(1):182–191
8. Furber SB, Galluppi F, Temple S, Plana LA (2014) The spinnaker project. Proc IEEE 102(5):652–665
9. Park CH, Kim SB (2015) Sequential random k-nearest neighbor feature selection for high-dimensional data. Expert Syst Appl 42(5):2336–2342

Investigation on Two Dimensional Photonic Crystal Based All Optical Logic Gates

M. Selvakumari, K. Rama Prabha, S. Robinson, and A. Sadiq Batcha

Abstract In this paper, the proposed design of OR logic gate function are realized in a hexagonal lattice using two dimensional Photonic Crystal (PC). The OR gate is constructed by varying the rod radius in this design and by creating Y—shaped waveguide. The OR gate is proposed by the phenomenon of resonator. The design is created by two input waveguide and one output waveguide for the reference input and the output. The functional parameters of the logic gates are the response time, bit rate and contrast ratio are investigated through 2D Finite Difference Time Domain (FDTD) method. The calculated functional parameter such as response time, bit rate and contrast ratio is 0.6948 ps, 1.4392 Tbps and 38.18 dB for logic 1.

Keywords OR gate · Photonic crystal · Band gap. finite difference time domain · Plane wave expansion

1 Introduction

The increasing growth of technology is leads to higher the demand for future optical communication. The optical telecommunication [1] system uses a transmitter to encode a message into optical signal and it carries the signal from transmitter to the receiver. The optical receiver decodes the received signal into message signal. In the optical devices and the networks are used to transmitting the signal and processing the higher performance [2] of the optical signal [3]. And theses method can achieve the higher bandwidth and high speed of optical signal. Optical communication is required to enhance the speed of data transmission and the optical signal received with high quality without degradation of the signal. But the electronic devices are not sufficient to transfer signal because the speed of operation due to the electro optical conversion

M. Selvakumari · K. R. Prabha · S. Robinson
Department of Electronics and Communication Engineering, Mount Zion College of Engineering and Technology, Pudukkottai, TamilNadu, India

A. S. Batcha (✉)
Faculty of Engineering & Built Environment, MAHSA University, Jenjarom, Malaysia

Z. Zakaria and S. S. Emamian (eds.), *Recent Advances in Electrical and Electronic Engineering and Computer Science*, Lecture Notes in Electrical Engineering 865, https://doi.org/10.1007/978-981-16-9781-4_14

method. The optical signal processing techniques overcomes the speed limitation and high power consumption. Now most of the designers are willing to design the optical devices in photonic crystal (PC). The PC has some special characteristics like low power consumption, high speed, compactness and it is flexible to design an integrated device [4].

A PC is a periodic optical nanostructure that affects the motion of photons in such a way that ionic lattice effect electrons in solids [5]. In the photons that propagates through the structure depends on the wavelength. The wavelength that propagates called modes and the group of allowed modes from bands and the disallowed bands of wavelength are called photonic band gaps. The optical structure can control the flow of light. Typically a two materials that are, dielectric and air are formed from the periodic modulation of their dielectric constant based on the structure [6]. In the PC can fabricated using three methods that are, one, two and three dimensional. One dimension (1D) is a simply a layered structure such as dielectric mirror, which may have a band gap for light entering the structure perpendicular to the layers. Two dimension (2D) PC layers there have been significant improvements in silicon based system and the improvements in integration with electronic circuits [7]. Further improvements can be expected in combined optical and electronic circuits with features such as switching, tuning, and delay functionality. Three dimensional (3D) PC have a sufficiently high refractive index, a suitable duration and a conductive filling rate that the entire band gap can be formed in the structure. In most of the designers can used 2DPC for optical design because it is easy to fabricate the electronic devices to optical devices. In the PC based components are logic gate [1–11], resonator [9–14], encoder [14, 15], flip flop [16], multiplexer [17], filter [18–21], adder [22], sensor [23–28].

Through there are several attempts are design to realize all-optical logic gates and it is limited by output transmission and complexity. In this design, all optical OR logic gate are designed using 2DPC in hexagonal lattice. And to calculate the functional parameter from the design such as, response time, bit rate and contrast ratio [8]. The plane wave expansion (PWE) method is used to calculate the PBG.

The photonic band gap mainly focused from band diagram, because this is important part of photonic crystal. A PBG crystal is a structure that can handle beams of light in the way of semiconductor control electronic currents [7]. In the semiconductor cannot support lying of energy in the electronic band gap. Similarly PC cannot support photons lying in the photonic band gap. By allowing light signal to propagate through a crystal, the light signal processing can be done by applying defects. In the 2DPC have two types of lattice structure such as, hexagonal lattice structure and square lattice structure [7].

In the below section are related to structure design of 2 and 3 input OR gate. The Sects. 2.1 and 2.2 include the two input and three input OR gate design, field distribution, output response and functional parameters. The Sect. 3 is based on the conclusion of the OR gate functional parameters such as contrast ratio, response time, bit rate in all optical logic gates.

2 Structural Design of OR Gate

The proposed design of OR logic gate consists of an array of 21*21 Si rods arranged on a hexagonal lattice and the truth table in below the Table 1 defines the logic of OR gate. Performance of logic gate is simulated using FDTD method. The band diagram is obtained the lattice constant of 600 nm, the rod radius is 120 nm and the refractive index is 3.46. The band diagram gives the propagation modes and the PBG of the proposed design (Fig. 1).

2.1 *Two Input OR Gate*

The proposed structure consists of two inputs and one output. The input port is port A and port B and the output port is port Y. OR gate is on hexagonal lattice of 21*21 rods on silicon material. The radius of rod in the whole structure is 120 nm and the refractive index is 3.46. The distance between the neighbor rods is 600 nm. Two input waveguides are created by removing 8 rods in a row. The output waveguide are created by reduce the radius (30 nm) of 7 rods in a row. The hexagonal lattice created by removing 6 rods in lift and right side of the design and to reduce rode radius (30 nm) in top and bottom of the hexagonal design. The band diagram shown below Fig. 2 has both TE PBG and TM PBG where the blue color indicates TE PBG and red color represents TM PBG. The first TE PBG is obtained between 0.5492 nm $< (a/\lambda) < 0.540$ nm whose corresponding wavelength range. The schematic structure of OR gate can be given below Fig. 3a and the 3Dimensional view of OR gate design can be given below Fig. 3d.

Table 1 Truth table for OR gate

Input A	Input B	Output Y
0	0	0
0	1	1
1	0	1
1	1	1

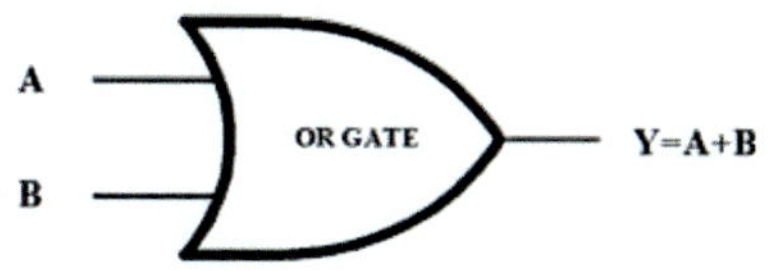

Fig. 1 Symbol of OR gate

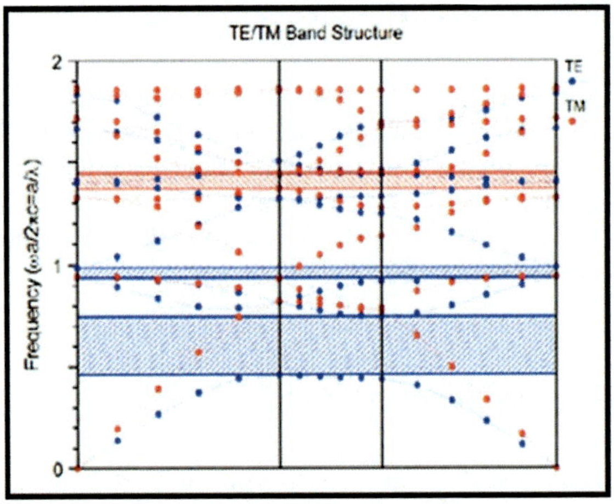

Fig. 2 Band diagram of the proposed structure design without defects

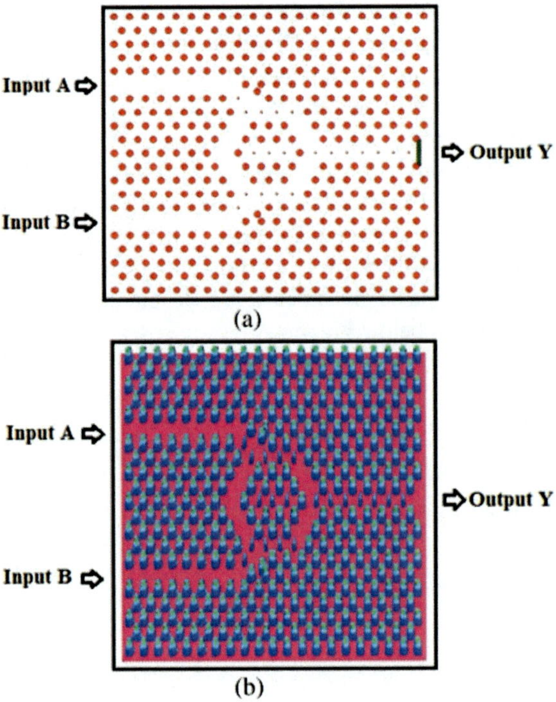

Fig. 3 **a** The schematic structure of two input OR gate and **b** 3D view of proposed design

The contrast ratio is one of the important parameters for designing the optical logic gates. The contrast ratio is obtained from minimum ON to OFF logic level of pro- posed logic gates. The contrast ratio is denoted a

$$CR = 10 \log PON/POFF$$

where, P_1 is the power level at logic 1 and P_0 is the power level at logic 0.

The field distribution of four logic of OR gate is shown in Fig. 4a–d. When no input power is launched to the input ports A and B, there is no output at port Y and the logic is considered as '0'. When the input A or B is launched to the input ports and e signal is propagated to reaches the output port Y and the logic is considered as '1'. When the two inputs A and B both are launched to the input port and the signal reaches the output port Y and the logic is also considered as '1'.

The output response of the OR gate is shown in above Fig. 5 Which calculate proposed the functional parameters such as contrast ratio, response time and bit rate in the proposed design. The contrast ration can be calculated as 38.18 dB. The average output power is calculated that the time required traveling the output power from 0 to 90% output. The time consists of two components namely t and t_1 where t is the transmission delay to reach 0.1% of P_{avg} and t_1 is the transmission delay to reach 90% of P_{avg}. For logic 01 and logic 10, the time $t = 1.2152$ ps and the time $t_1 = 1.3889$ ps. From that the response period of proposed two input OR gate is accounted as 0.6948 ps and it operates at the bit rate of 1.4392 Tbps (Table 2).

The impact of functional parameters of the device, namely, output power, response time, contrast ratio and bit rate while varying the inner defect rod(R) radius are investigated which are depicted in Table 3. Though there are several attempts are made by varying the inner rod radius at 30 nm which attains minimum response time. The contrast ratio is increasing linearly while increasing the inner rod radius. However, the response time is reduced significantly. Finally, Fig. 6 shows the effect of bit rate and response time for different inner rod radius. However, the finest functional parameters, such as, bit rate of 1.4392 Tbps and response time of 0.694 ps are arrived at 30 nm of inner rod radius, hence, it is considered for this attempt.

2.2 *Three Input OR Gate*

The three input OR gate symbol and truth table can be shown above Fig. 7 and Table 3. The proposed design of three input OR gate is on hexagonal lattice of 21*21 rods on silicon material. The radius of rod in the whole structure is 120 nm and the refractive index is 3.46. The distance between the neighbor rods is 600 nm. Two input waveguides are created by removing 8 rods in a row and one reference input waveguide are created by removing 7 rods in the row. The output waveguide are created by reduce the radius (30 nm) of 7 rods in a row. The hexagonal lattice created by removing 6 rods in lift and right side of the design and to reduce rode radius

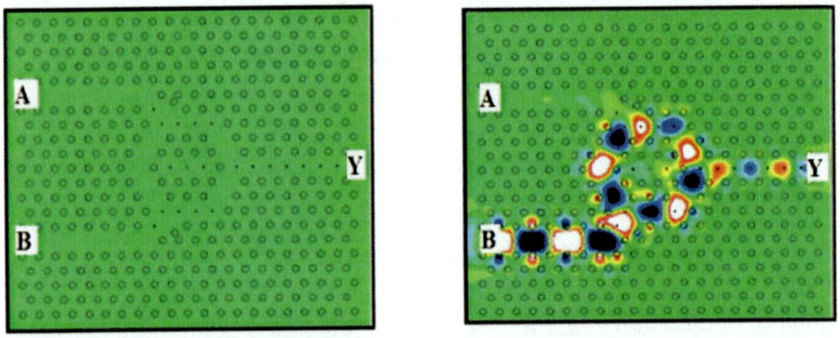

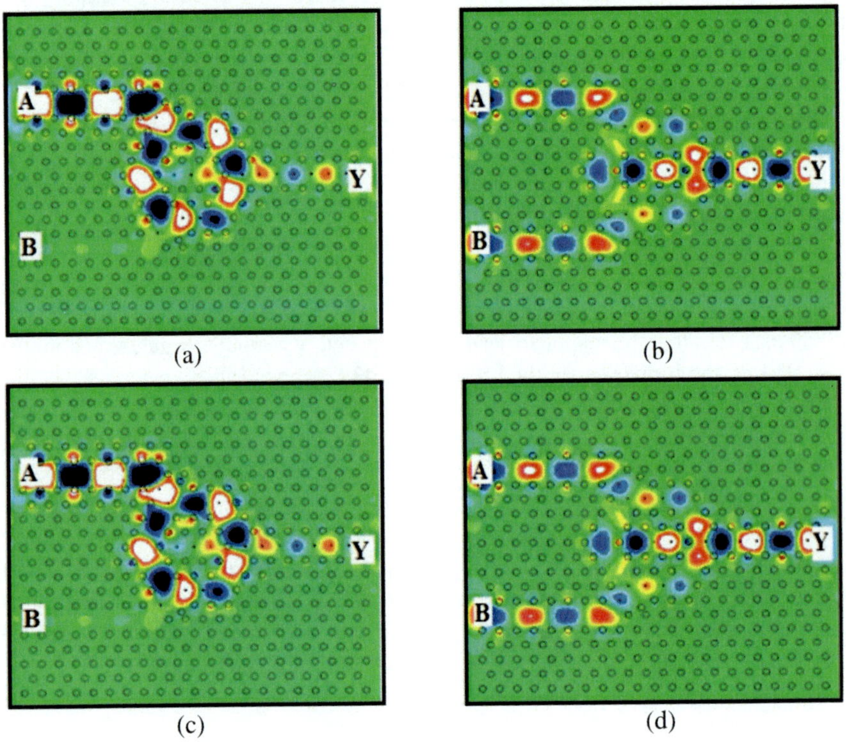

Fig. 4 Field distribution of two input OR gates **a** A = 0, B = 0 and Y = 0, **b** A = 0, B = 1 and Y = 1, **c** A = 1, B = 0 and Y−1 and **d** A = 1, B = 1 and Y = 1

(30 nm) in top and bottom of the hexagonal design. The schematic structure of OR gate can be given below Fig. 8a and the 3Dimensional view of OR gate design can be given below Fig. 8b.

The field distribution of four logic of OR gate is shown in Fig. 9a–d. When 001 input power is launched to the input ports A and B, and reference input port C there is no output at port Y and the logic is considered as ‘0’ shown above Fig. 9a. When the input A or B and reference input C is launched to the input ports and the signal

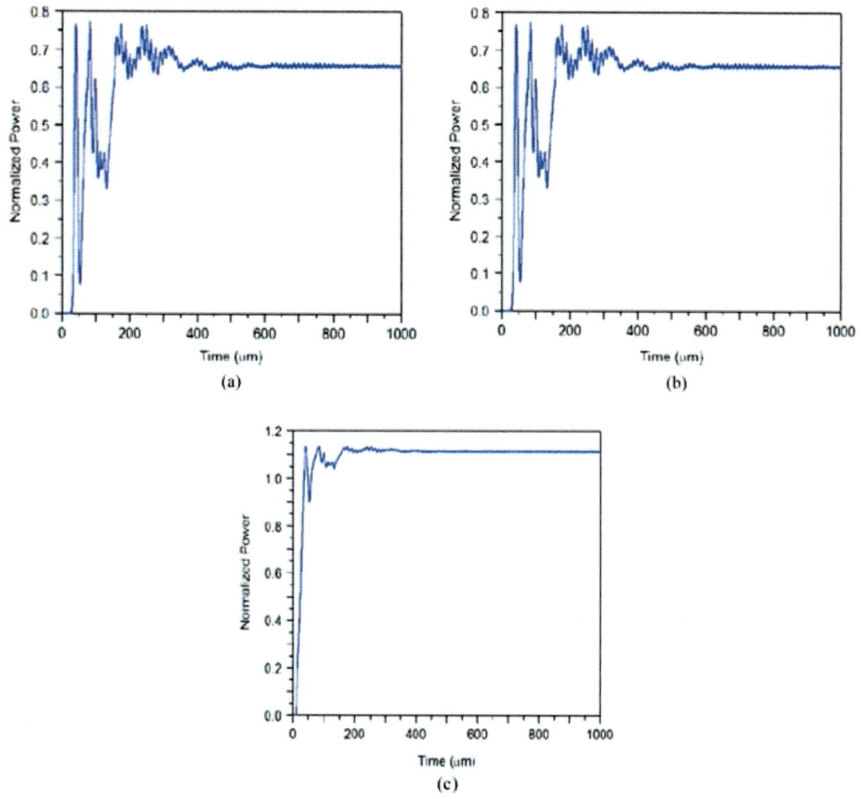

Fig. 5 Output response of two input logic gate **a** logic 01 **b** logic 10 and **c** logic 11

Table 2 Truth table and functional parameter of the two input OR gate

Input A	Input B	Output Y	Contrast ratio (dB)	Response time (ps)	Bit rate (Tbps)
0	1	1	38.18	0.6948	1.4392
1	0	1			
1	1	1			

is propagated to reaches the output port Y and the logic is considered as '1' shown above Fig. 9a–c. When the three inputs port A, B and C these signals are launched to the input port and the signal reaches the output port Y and the logic is also considered as '1' shown above Fig. 9d.

The output response of three OR gate is shown in above Fig. 10a–c. Which calculate proposed the functional parameters such as contrast ratio, response time and bit rate in the proposed design. The functional parameter can be tabled in below Table 4. The contrast ration can be calculated as 40 dB. The average output power is calculated that the time required traveling the output power from 0 to 90% output.

Table 3 Different functional parameter of two input OR gate

Radius (nm)	Contrast ratio (dB)	Response time (ps)	Bit rate (Tbps)
10	39.37	1.3889	0.7199
20	39.61	1.3889	0.7199
40	39.40	1.5984	0.6256
50	39.60	5.3242	0.1872
60	39.21	17.5924	0.0568
70	40	5.7873	0.1724
80	39	7.1761	0.1393
90	39.02	1.8518	0.54001

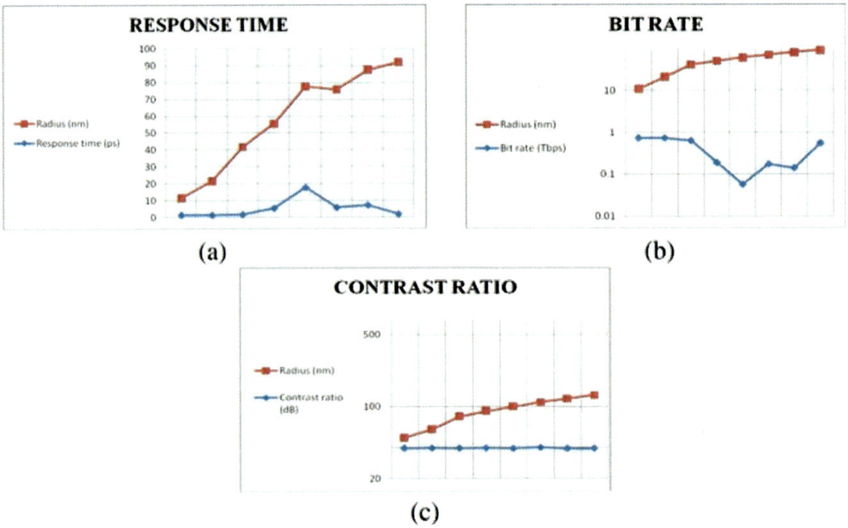

Fig. 6 Output response of functional parameters

Fig. 7 Symbol of three input OR gate

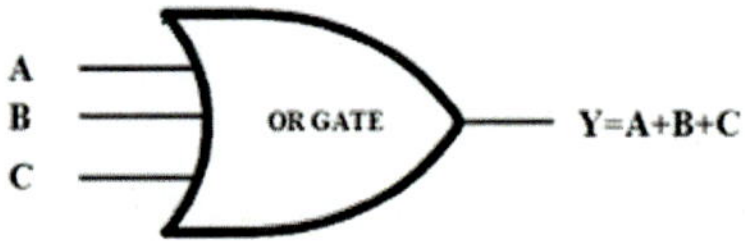

The time consists of two components namely t and t_1 where t is the transmission delay to reach 0.1% of P_{avg} and t_1 is the transmission delay to reach 90% of P_{avg}. For logic 011, logic 101 and logic 111, the time t = 0.365 ps and the time t_1 = 0.964 ps. From that the response period of proposed two input OR gate is accounted as 2.356 ps and it operates at the bit rate of 0.424 Tbps (Tables 5 and 6).

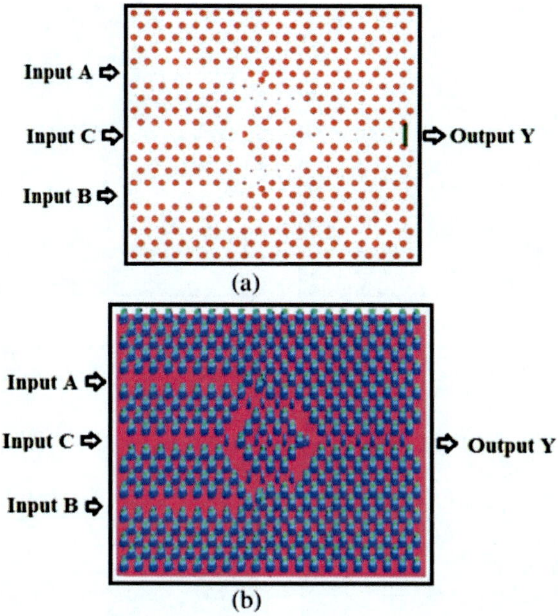

Fig. 8 **a** The schematic structure of three input OR gate and **b** 3D view of proposed design

From the above comparison Table 7 it is clearly noted that the proposed work which offers better response time and high contrast ratio. The impact of functional parameters of the device, namely, output power, response time, contrast ratio and bit rate while varying the inner defect rod(R) radius are investigated. The response time is increased and the bit rate is decreased, if the inner rod radius is increased above 20 nm. However, the functional parameters, such as, bit rate of 0.424 Tbps and response time of 2.356 ps are arrived at 20 nm of inner rod radius. The proposed structure is compact. Alternatively, if the response time is better, it limits the contrast ratio. In addition, if the contrast ratio, bit rate and response time are better, the size of the OR gate is larger. However, the proposed OR gate provides significant improvement in all the parameters with the compact size than the reported logic gates. Hence, the proposed encoders are used for high speed photonic integrate circuits and switching networks.

3 Conclusions

The proposed design of two input and thee input logic gate are based on the 2DPC and the functional parameter are constant ratio, response time and bit rate. The functional parameters are calculated using FDTD method. The wavelengths are operated at 1550 nm. The calculated parameters such as contrast ratio, response time and the

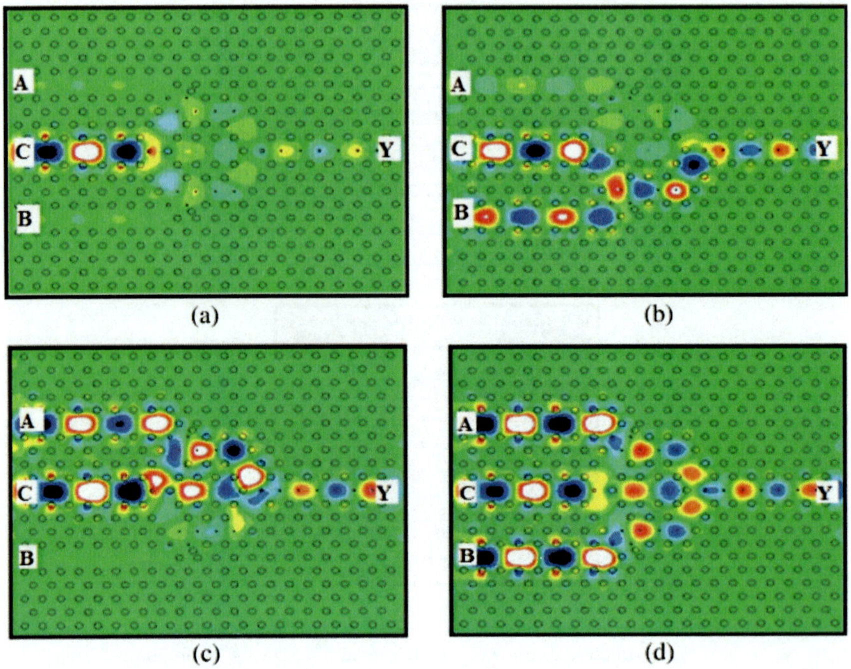

Fig. 9 Field distribution of proposed three input OR gates **a** A = 0, B = 0, C = 1 and Y = 0, **b** A = 0, B = 1, C = 1 and Y = 1, **c** A = 1, B = 0, C = 1 and Y = 1 and **d** A = 1, B = 1, C = 1 and Y = 1

bit rate are 38.18 dB, 0.6948 ps and 1.4392 Tbps respectively. The three input OR gate calculated parameters such as contrast ratio, response time and the bit rate are 40 dB, 2.356 ps and 0.424 Tbps respectively. The compact structure of proposed two input OR logic gate and three input OR gate which offers high bit rate, better signal operating speed and achieve high contrast ratio. Hence the proposed structure of OR logic design is suitable for designing high speed optical communication devices and high speed switching device.

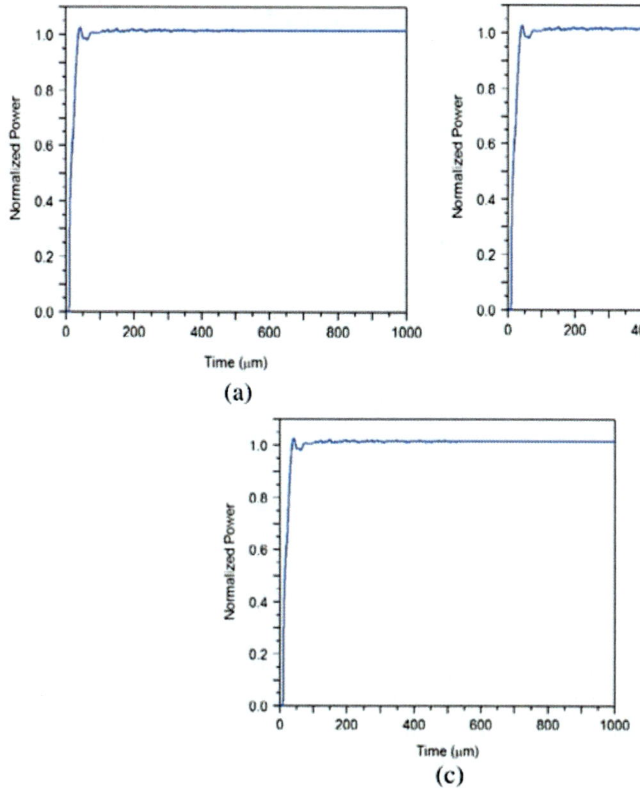

Fig. 10 Output response of two input logic gate **a** logic 011 **b** logic 101 and **c** logic 111

Table 4 Truth table for three input OR gate

Input A	Input B	Input C	Output Y
0	0	0	0
0	0	1	1
0	1	0	1
0	1	1	1
1	0	0	1
1	0	1	1
1	1	0	1
1	1	1	1

Table 5 Truth table and functional parameter of the two input OR gate

Input A	Input B	Output Y	Contrast ratio (dB)	Response time (ps)	Bit rate (Tbps)
0	1	1	40 dB	2.356 ps	0.424Tbps
1	0	1			
1	1	1			

Table 6 Different functional parameter of two input OR gate with respect to the radius

Radius (nm)	Contrast Ratio (dB)	Response Time (ps)	Bit Rate (Tbps)
10	39.67	0.4644	2.1533
30	39.70	1.1165	0.8954
40	40.78	0.6701	1.4922
50	40.82	0.8933	1.1194
60	40.67	0.9244	1.0817
70	40.48	8.7972	0.1136
80	41.20	13.898	0.0719
90	41.97	3.4724	0.2879

Table 7 Existing system functional parameter for two and three input OR gate

References	Lattice	Defect	Wavelength	Size μm^2	Contrast ratio (dB)	Response time (ps)	Bit rate (Tbps)	
Jayson et al. [7]	Hexagonal 2 input OR gate	Y Shape	1226 μm	12.6× 11	30	0.7846	1.27	
	Hexagonal 3 input OR gate	Y Shape	1226 μm	12.6 × 11	30	0.52	1.92	
Ishizaka et al. [9]	Square	Y Shape	1550 nm	***	10.79	0.9	1.1	
Pirzadi et al. [17]	Square	***	1550 nm	***	6.767	0.4	2.5	
Robinson and Shanthi [13]	Square	Y Shape	1550 nm	***	18.7	3	0.333	
Mehdizadeh et al. [18]	Hexagonal	Y shape	***	12.5× 16	7.27	***	0.8	
	Proposed work	Two input OR gate hexagonal	Y shape	1550 nm	21×21	38.18	0.6948	1.439
		Three input OR gate Hexagonal	Y shape	1550 nm	21×21	40	2.356	0.424

References

1. Younis RM, Areed NFF, Obayya SSA (2014) Fully integrated AND and OR optical logic gates. IEEE Photon Technol Lett 26(19):1900–1903
2. Rani P, Fatima S, Karla Y (2017) Realization of all optical logic gates using universal NAND gates on photonic crystal platform. Super Lattices Microstruct 109:619–625
3. Ajaykumar SK, Raghuwan S (2014) Implementation of XOR/XNOR and AND logic gates by using Mach-Zehnder interferometer. Optik Int J Light 125(19):5764–5767
4. Shaik EH, Rangaswamy N (2015) Design of photonic crystal based all optical AND gate using T shaped waveguide. J Modern Optics 63(10):941–949
5. Rani P, Kalra Y, Sinha RK (2013) Realization of AND gate in photonic crystal Waveguide. Opt Commun 298–299:227–231
6. Shaik EH, Rangaswamy N (2016) Improved design of all-optical photonic crystal logic gates using T-shaped waveguide. Opt Quantum Electron 48(3):335
7. Jayson KJ, Subhalakshmi G, Robinson S (2018) Performance evaluation of two dimensional photonic crystal based all optical AND/OR logic gate. J Opt Commun 39(3)
8. Saranya D, Anbazhagan R (2020) Design and analysis of optical logic gates based on trifurcation structured 2D photonic crystals. Opt Quantum Electron 32(2)
9. Ishizaka Y, Kawaguchi Y, Saitoh K, Koshiba M (2011) Design of ultra- compact all-optical XOR and AND logic gates with low power consumption. Opt Commun 284(14):3528–3533
10. Salmanpour A, Mohammadnejad S, Bahrami A (2015) All optical photonic crystal AND, XOR, and OR logic gates using nonlinear Kerr Effect and Ring resonators. J Mod Opt 62(9):693–700
11. Ashkan P, Zavvari M, Alipour-Banaei H (2015) All-optical AND/OR/NOT logic gates based on photonic crystal ring resonators. Front Optoelectron 9(4):578–5843
12. Salmanpour A, Mohammadnejad S, Omran PT (2015) All optical photonic crystal NOT and OR logic gates using nonlinear Kerr effect and ring resonators. Opt Quant Electron 47(12):3689–3703
13. Robinson S, Shanthi KV (2016) Analysis of protein concentration based on photonic crystal ring resonator. Int J Opt Photonics 10:123–3
14. Suganya T, Robinson S (2017) 2D Photonic crystal based biosensor using rhombic ring resonator for glucose monitoring. ICTACT J Microelectron 3:349–400
15. Salimzadeh A, Alipour-Banaei H (2018) An all optical 8 to 3 encoder based on photonic crystal OR-gate ring resonator. Opt Commun 410:793–798
16. Goudarzi K, Mir A, Chaharmahali I et al (2016) All optical XOR and OR logic gates based on line and point defects in 2-D photonic crystal. Opt Laser Technol 78:139–142
17. Pirzadi M, Mir A, Bodaghi D (2016) Realization of ultra-accurate and compact all-optical photonic crystal OR logic gate. IEEE Photonics Technol Lett 282(1):2387–2390
18. Mehdizadeh F, Sooroosh M, Banaei HA (2017) Proposal for 4-to- 2 encoder based on photonic crystals. IET Optoelectron 11:29–36
19. Moniem TA (2015) All-optical S-Flip-flop using 2-D photonic crystal. Optical Quantum Electron 45:2843–2848
20. Venkatachalam K, Robinson S, Dhamodharan SK et al (2017) Performance analysis of an eight channel demultiplexers using a 2D photonic crystal quasi square ring resonator. Opto-Electron Rev 25(2):74–79
21. Robinson S, Nakkeeran R (2013) Investigation of parameters affecting the performance of two dimensional photonic crystals based pass band filter. Optical Quantum Electron 43:1–14
22. Li Z, Liu Y, Zhang S, Ju H et al (2005) All- optical logic gates based on an soa and an optical filter. 31st European Conference on optical communication 2: 229
23. Robinson S, Nakkeeran R (2011) Investigation on two dimensional photonic crystals resonant cavity based band pass filter. Optik Int J Light Optics 123(5):451–457
24. Stubkjaer KE (2000) Semiconductor optical amplifier based all-optical gates for high-speed optical processing. IEEE J Sel Top Quantum Electron 6(6):1428–1435
25. Serajmohammad S, Alipour H, Mehindeh F (2018) Proposal for realizing an all-optical half adder based on photonic crystals. Appl Opt 57:1617–1624

26. Shanthi KV, Robinson S (2014) Two dimensional photonic crystal based sensor for pressure sensing. Photonic Sens 4(3): 248–253
27. Singh BR, Rawal S (2015) Photonic crystal based all optical NOT logic gate. J Opt Soc Am 32(12):2260–2263
28. Arunkumar R, Suganya T, Robinson S (2017) Design and analysis of photonic crystal elliptical ring resonator based pressure sensor. Photonics Opt Technol 3(1):30–34

Miniaturized Multiband Metamaterial Antenna for 5G Applications

Check for updates

V. Ramya, B. Vanniya, S. Robinson, and A. Sadiq Batcha

Abstract A left handed metamaterial antenna is designed in this paper. The unit cell is a combination of Double sided Coupled Split Ring Resonator (DSRR) and Capacitance Loaded Strips (CLS) in order to achieve a design that exhibits both negative magnetic permeability and negative electrical permittivity which has a capability of achieving negative refraction to strengthen the radiation power of the antenna. The antenna is designed using HFSS software and it is etched on a FR4 substrate having a thickness of 0.8 mm and the overall dimension of the antenna is $13 \times 13 \times 0.8$ mm^3.The antenna achieves LHM multiband frequency for a measured return loss whose center frequencies are 4.7 GHz, 7.6 GHz, 8.8 GHz, 10 GHz, 10.5 GHz and measured voltage standing wave ratio are 1.12 dB, 1.76 dB, 1.097 dB, 1.25 dB, 1.55db with a maximum gain and directivity of 2.1 dB and 2.9 dB respectively. The efficiency of the proposed antenna is 95% at the frequency of 8.8 GHz which is considerable higher. The proposed metamaterial antenna is better miniaturized, and has multiband frequencies which mainly find applications in various wireless communication systems such as WI-MAX, fixed mobile and satellite communication systems.

1 Introduction

Negative—Index metamaterial or Negative—Index Material (NIM) is a metamaterial whose refractive index for an electromagnetic wave has a negative value over a frequency range [1]. The metamaterial having a negative refractive index n is called left handed material. In recent times, there have been studies on these material

V. Ramya · S. Robinson
Department of Electronics and Communication Engineering, Mount Zion College of Engineering and Technology, Pudukkottai, TamilNadu, India
e-mail: ramyaharshada@gmail.com

B. Vanniya
Alagappa Chettiar College of Engineering and Technology, Karaikudi, TamilNadu, India

A. S. Batcha (✉)
Faculty of Engineering & Built Environment, MAHSA University, Jenjarom, Malaysia

Z. Zakaria and S. S. Emamian (eds.), *Recent Advances in Electrical and Electronic Engineering and Computer Science*, Lecture Notes in Electrical Engineering 865, https://doi.org/10.1007/978-981-16-9781-4_15

and theoretically discussed by Veselago. NIMs are constructed by basic part called unit cells which are significantly smaller than the applied electromagnetic radiation. Varieties of metamaterial have been envisioned and fabricated [2, 3]. In this paper, a multiband left handed metamaterial (LHM) antenna is proposed based on the working principle [4–6]. The constructions of unit cells are made by arranging wires periodically [7], by providing a broadside Coupled Split Ring Resonator can exhibit an effective negative refractive index [8, 9]. Based on the complete analysis from [10], it was established that the Broadside Coupled SRR (BS - SRR) had the improvement of having a much smaller electrical size as compared to the conventional SRR [11]. In most recent times, lots of studies have been lead to examine the performance of micro strip patch antenna in the existence of left handed material.

Suresh et al. was designed a multiple parasitic patches antenna with the overall dimensions of $14 \times 13.4 \times 1.6\ \text{mm}^3$ and the result reports that the antenna radiates at 2.44 GHz mainly used for wideband wireless applications. Anshuman et al. uses a Moth—Flame optimization algorithm to present the microstrip patch antenna (MPA) with the overall dimensions of $50 \times 50\ \text{mm}^2$. The result reported that the operational bandwidth was 3.1 GHZ and the return loss was −20 dB that covers UWB applications. Jenath et al. use FRDGS design for multi standard wireless communication systems. The results report that the antenna achieved UWB over a frequency ranges from 1.33 to 8.7 GHz and obtain maximum gain of 3.19 dB. Wen-Chung et al. composed dipole antennas was directly connected to three of six LHM unit cells and are arranged in a 2×3 antenna array form. The maximum gain and directivity was reported as −1dBi and 3dBi, respectively, the return loss measurements demonstrated that this antenna achieves 63% bandwidth and with 40% radiation efficiency at 2.5 GHz. Mehedi Hasan et al. designed a left handed metamaterial (LHM) antenna was developed with an overall dimension of $25 \times 18\ \text{mm}^2$ for Wi-Fi, WLAN and WiMAX applications. The measured frequency band includes 2.36 GHz to 2.49 GHz and 3.07 GHz to 4.0 GHz and the average gain of 1.87 dBi and covers bandwidth of about 130 MHz and 930 MHz respectively. Sourav Nandi et al. proposed a Triple band MIMO antenna for WLAN and WiMAX applications. The result reports that the proposed antenna resonates at 2.4 GHz and 3.5 GHz with return loss of −26 dB and −22 dB and maximum gain of −2 dBi and 0.14 dBi, respectively. Abed et al. proposed a single layer slot antenna used for Wi-Fi and WiMAX applications. The return loss measurement demonstrated that the antenna achieved a VSWR less than 2 dB at resonating frequency of 2.45 and 3.6 GHz with maximum gain of 1.85 dBi. Rengasamy et al. developed a compact open split ring resonator (COSRR) metamaterial and the result reports that the antenna radiates at different frequencies of 2.4, 5.2, 5.5, 5.8 and 7.4 GHz with a maximum gain of 0.37, 1.61, 1.88, 2.08, and 2.52 dB respectively. Mehri et al. proposed a microstrip patch antenna (MPA) with DGS for UWB applications with overall dimension of $20 \times 20\ \text{mm}^2$. The antenna resonates at 3–13.6 GHz for a VSWR less than 2 with maximum gain of 0–6 dB.

From the review on previous presented papers in [12–20], were good for 5G network, but not suitable for a 5G portable device because of bulky configuration large size and negative gain obtained.

To overcome the problems, Double side split ring resonator (DSSR) with capacitance loaded strips (CLS) is presented in this paper. This design antenna is small in size then the previously presented paper. Furthermore, the proposed antenna provides good gain and radiations characteristics for multiband frequencies.

Section 1 support the literature survey to the work proposed in this paper. Section 2 details the left handed metamaterial (LHM) multiband frequency antenna design for the requirements of multiband of 5G requirements used for Wi-Max, mobile communication and satellite communication systems. Section 3 details the metamaterial unit cell structure and also details the retrieving the constitutive effective parameters. Section 4 details the proposed antenna measured results and compares the previously presented papers results with the proposed antenna results. Section 5 concludes the work proposed are found to be satisfying the mid rand of 5G requirement.

2 Antenna Design and Configuration

In this section, geometry of proposed antenna is shown in Fig. 1 and Table 1 provides the details about the geometrical parameters of the proposed antenna. The antenna is printed on both the side of a low cost FR4 substrate with thickness 0.8 mm and the relative permittivity of 4.4. The antenna structure consists of a combination of rectangular SRRs (RSRR) and the CLSs as shown in the Fig. 2, which is united into a single line feeling strip altogether deposited on the top surface of the substrate.

Figure 1a shows an ordinary planar patch antenna which provides a bandwidth of 7.9–9 GHz. Width and length of antenna are represented as l_2 and l_3. The formulas given below (1, 2) are used to design a microstrip patch antenna (MPA) [21]. Ground is provided at the back side of the substrate. Width and length of the ground are represented as l_g and W_g. According to Fig. 1b, in order to improve the impedance bandwidth and to obtain multiband frequencies, CLS is loaded on a patch achieves

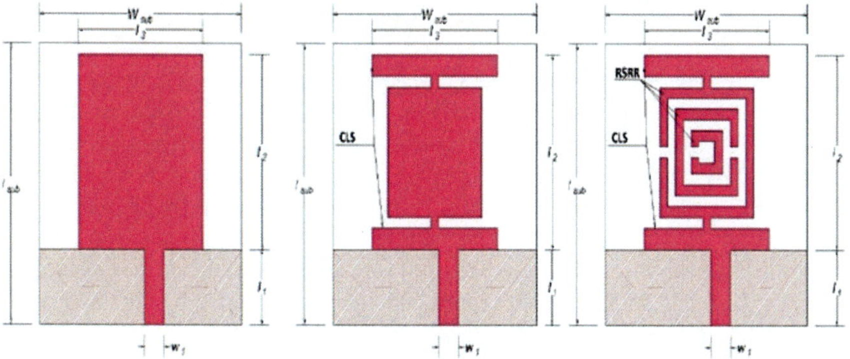

(a) Planar antenna (b) CLS loaded patch (c) CLS and RSRR loaded patch

Fig. 1 Simulated various shaped antenna

Table 1 Dimension of antenna parameters (with respect to Fig. 2)

Parameter	Dimensions (mm)
l_1	3.5
l_2	9
l_3	8
l_g	3.5
l_{wire}	10
W_1	1.25
W_2	0.25
W_g	13
W_{sub} x l_{sub}	13 × 13

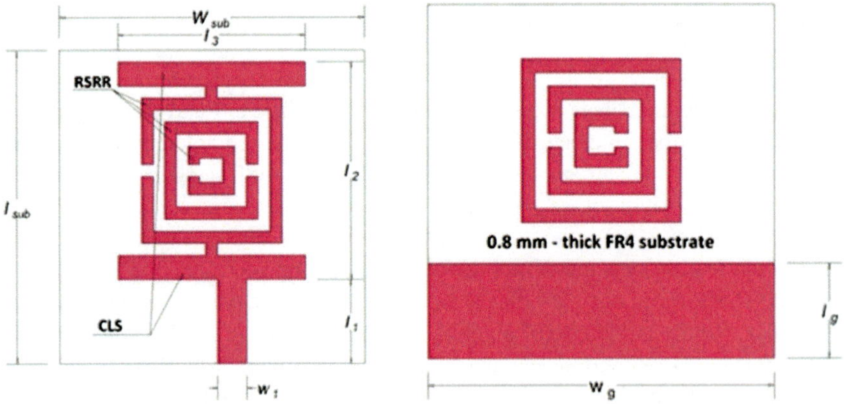

(a) RSRR and CLS loaded patch (Top view)

(b) Ground added with Inverted RSRR (Back View)

Fig. 2 Geomentry of proposed antenna

dual band resonates at 6.7 and 10.5 GHz and also provides a bandwidth of 6.2–7.2 GHz and 10.2–10.7 GHz. As show in Fig. 1c, by adding a broad side coupled split ring resonator (BC-SRR) on the patch, antenna radiates at five different frequencies, 3.5, 6.8, 9.2, 10.3 and 10.7 GHz used for Wi-Fi applications, mobile communications and satellite communications. From Fig. 1c, bandwidth obtained are not seems to be expected.

Figure 2, shows a front and back view of proposed antenna. On the top side of the substrate, CLS and RSRR are introduced and antenna is excited by 50Ω microstrip feed line by an SMA connector. And at the bottom side of the substrate inverted version of RSRR is added. The results reports that the proposed antenna achieves multiband LHM antenna which radiates at 4.7, 7.6, 8.8, 10 and 10.5 GHz. And also achieves UWB frequency range of 9.8 GHz to 10.8 GHz.

Width of the patch is

$$l_2 = \frac{c}{2fr}\sqrt{\frac{2}{\varepsilon r + 1}} \tag{1}$$

where,

c is the velocity of light in free space,

f_r is the resonance frequency,

ε_r is the dielectric constant of the substrate.

Patch Length is

$$l_3 = \text{Leff} - \Delta\text{L} \tag{2}$$

Effective Length

$$\text{Leff} = \frac{\varepsilon r + 1}{2} + \frac{\varepsilon r - 1}{2}\frac{1}{\sqrt{1 + 12\frac{h}{Wp}}} \tag{3}$$

Extension in length

$$\Delta\text{L} = 0.412h\frac{(\varepsilon\text{ref} + 0.3)\left(\frac{Wp}{h} + 0.264\right)}{(\varepsilon\text{ref} - 0.258)\left(\frac{Wp}{h} + 0.8\right)} \tag{4}$$

where h is the height of the substrate.

The return loss and VSWR with respect to the frequency is shown in Fig. 3 and Fig. 4, respectively. As shown in Figs. 3 and 4, by adding CLS and RSRR on the top side of the substrate and by adding inverted RSRR on the bottom surface of the substrate along with ground the presented antenna satisfies the multiband frequency with improved bandwidth (Table 2).

3 Metamaterial Unit Cells Configuration

3.1 Descriptions of Unit Cell Structures

From the above mentioned literature review, the most well-known method of left handed metamaterial design is the combination of SRRs and CLS [22, 23]. In this paper the combination of rectangular SRRs (RSRR) and the broad—side coupled SRRs (BC-SRR) to form double sided coupled SRR is studied.

In this design, both negative μ and negative ε are achieved.

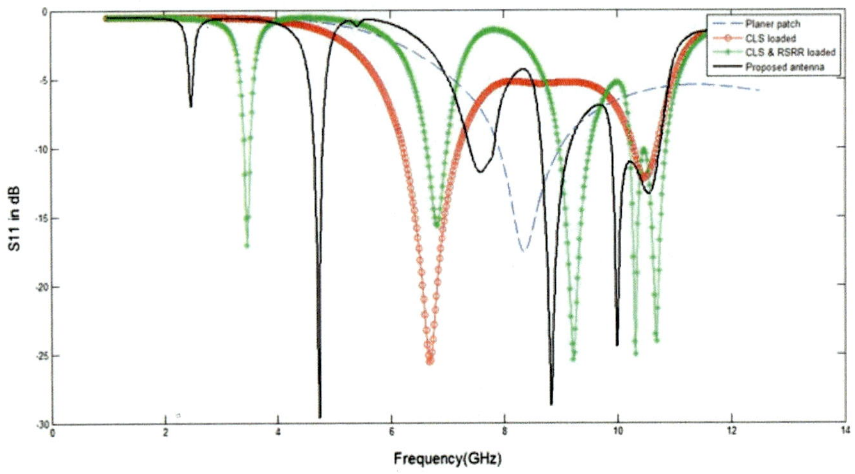

Fig. 3 Simulated return loss of various structured antennas

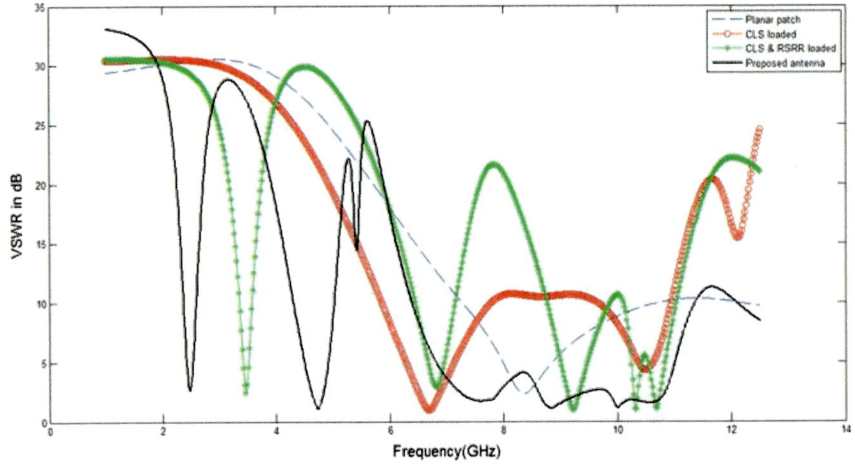

Fig. 4 Simulated VSWR of various structured antennas

Using full wave electromagnetic examination, transmission characteristics of the SRR and DSRR are explored to find out the advantage of this newly suggested magnetic resonator. The performance of LHM indicates smaller unit cell size compared to the free space wavelength (λ_0), and upper limit of the unit cell is one—fourth of the guided wavelength. The proposed LHM structures are shown in Fig. 5, with geometrical parameters given in the Table 3.

On the top surface of the substrate, the inner RSRR are surrounded by the outer RSRR, which is directly connected to CLS structure and have a DSRR attached to the bottom surface of the substrate. As a whole, this structure mainly purposes

Table 2 Comparison of various structured antennas with the proposed work

Type	Application bands covered (GHz)	Return loss, S11 (dB)	VSWR (dB)	Antenna type	Bandwidth (GHz)
Planar patch	8.4	−17.6	2.3	Single band	7.9–9
CLS loaded patch	6.7/10.5	−25.5/ −12.25	0.9/ 4.2	Dual band	6.2–7.2/ 10.2–10.7
CLS and RSRR loaded patch	3.5/6.8/9.2/10.3 /10.6	−17/ −15.5/ −25.4/ −25/ −24	2.4/ 2.8/ 0.93/ 0.97/ 1.08	Multi band	Narrow band at 3.5 GHz/ 6.6–7/ 9–9.5/ 10.2–10.8
Proposed antenna	4.7/ 7.6/ 8.8/ 1 0/ 10.5	−29.6/−11.7/ −28.7/ −24.4/ − 13.5	1.12/ 1.76/ 1.097/ 1.25/ 1.5	Multi band	Narrow band at 4.7 GHz/ 7.3–7.8/ 8.6–9.1/ WB at 9.8–10.8

to simplify and reduce the size of the structure and capable of producing a band of negative refractions indices at the microwave frequencies. An added advantage of this structure is entirely realizable and it is easy to assemble using conventional planar technology, such as microstrip.

The proposed LHM were simulated using FR4 epoxy substrate at a dielectric constant of 4.4 using High Frequency Simulation Software (HFSS) in order to obtain magnitude of S parameters in terms of return loss (S11) and insertion loss (S21). For simulation purpose, Unit cells were excited with an electric field along x—axis and magnetic field perpendicular to the conducting element along z—axis. For the proposed LHM unit cell, simulated S parameters has a transmission peaks at the frequencies of 3.6 and 7 GHz which indicates Left handed band. Thus the presence of a negative index band is expected for this structure, which could be determined by retrieving the constitutive effective parameters in the upcoming section. The magnitude of S parameters (for Fig. 3) is shown in Fig. 6.

3.2 Retrieving the Constitutive Effective Parameters

The electromagnetic properties of the LHM design is verified from retrieval algorithm described in [24–26]. It was used to obtain the constitutive effective parameters based on the reflection and transmission coefficient characteristics, Impedance z, the relative effective permittivity ε, the permeability μ, and the refractive index n. These are obtained by the following equations.

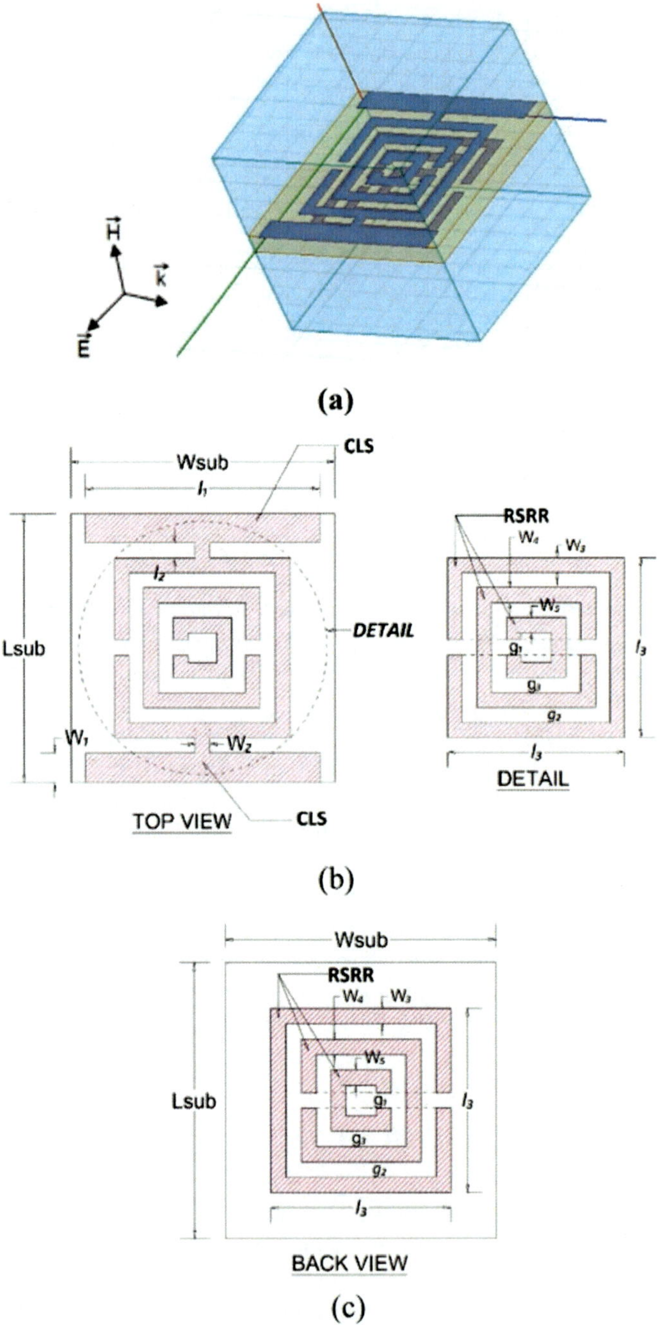

Fig. 5 Geometry of LHM unit cell consists of RSRR, CLS and DSRR. The DSRR producing negative ε is centered on the opposite side of the substrate: **a** Perspective view **b** top view and **c** back view

Table 3 LHM Unit cells design parameter for both Type I and Type II

Parameter	Dimensions (mm)
Unit cell size	9
Substrate thickness h	0.8
Length of CLS (l_1)	8
Distance between CLS and outer RSRR	0.5
Length of outer RSRR (l_3)	6
Width of CLS (W_1)	1
Width of the distance from CLS and outer RSRR (W_2)	0.5
Width of the outer RSRR (W_3)	0.5
Width of the center RSRR (W_4)	0.5
Width of the inner RSRR (W_5)	0.5
Width of wire strip (W_6)	0.25
Split length in the inner/outer RSRR (g_1)	0.5
Space between outer and center RSRR (g_2)	0.5
Space between center and inner RSRR (g_3)	0.5

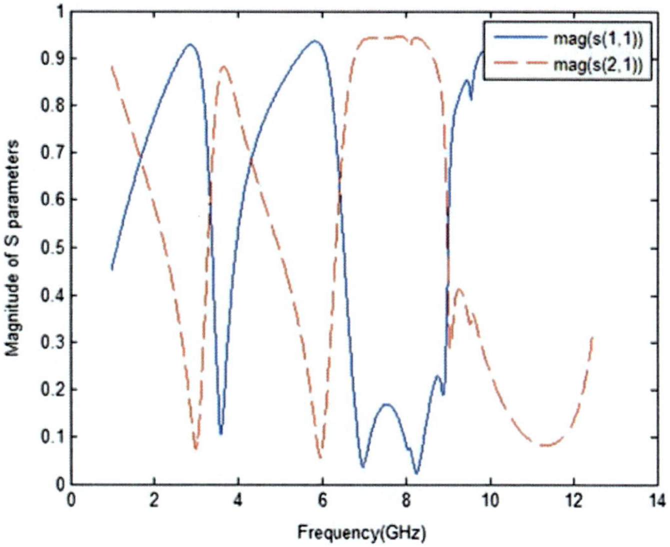

Fig. 6 Simulation results of S parameters unit cell in Fig. 3

$$z = \pm\sqrt{\frac{(1 + S11)^2 - S21^2}{(1 - S11)^2 - S21^2}} \tag{5}$$

$$e^{(\text{jnk}_0\text{d})} = \frac{\text{S21}}{1 - S11\frac{z-1'}{z+1}} \tag{6}$$

$$\text{n} = \frac{1}{k_0 d}\left[\left\{\left[\ln\left(e^{(\text{jnk}_0\text{d})}\right)\right] + 2m\pi\right\} - j\left[\ln\left(e^{(\text{jnk}_0\text{d})}\right)\right]'\right] \tag{7}$$

where, k_0 is the wave number, d is the maximum length of the unit element, m is the branch due to the periodicity of the sinusoidal function, (.)" represents the complex component and (.)' represents the real component of the complex number, E and H are the electric and magnetic field component, respectively.

The relationship between permittivity (ε), permeability (μ), refractive index (n) and the impedance (z) are given by the expression (8) and (9):

$$\varepsilon = \text{n} / \text{z} \tag{8}$$

$$\mu = \text{n x z} \tag{9}$$

Due to the multiple branches of the sinusoidal and square root function, there are multiple solutions for the refractive index and impedance. The unique solution can be approached by significant passive nature of the material and by properly selecting the branch m in Eq. (7) gap in the resulting refractive index can be avoided. In this paper, the largest dimension of the unit element which is less than one-sixth of wavelength in the material at 5 GHz is studies. This allow to take fundamental branch ($m = 0$) for a continuous refractive index. The retrieved effective parameters of proposed LHM structures are shown in the Fig. 7. In Table 4, negative refractive index frequency bands are summarized in detail.

Figure 8 illustrates Co polarization and cross polarization radiation characteristics of proposed LHM multiband antenna for different frequencies at 4.7, 7.6, 8.8, 10 and 10.5 GHz. Radiation characteristics are observed like dipole and omnidirectional radiation pattern in E—Plane (XY Plane) and H—Plane (YZ Plane).

Figure 9, shows the simulated current distributions of proposed multiband antenna at different frequencies (4.7, 7.6, 8.8, 10 and 10.5 GHz). Form Fig. 9a, the current distributions are maximum along ground plane and in the inner ring and minimum in the feed line and in outer ring of radiating element at 4.7 GHz. For 7.6 GHz, the maximum current distributions are along ground plane, in inner and in outer ring, but minimum concentration in the feed line.. For 8.8 GHz, the minimum current distributions are in the feed line and in the inner ring but maximum concentration in the outer ring, and along ground plane. From Fig. 9d, for 10 GHz the maximum concentration of current distributions are in inner ring, outer ring, CLS and along ground plane. For 10.5 GHz, minimum current distributions are in feed line and in outer ring.

The simulated results of proposed multiband antenna gain are measured as 1.2/ 1.3/ 2.1/ 1.2 and 1.4 dB at 4.7, 7.6, 8.8, 10 and 10.5 GHz. The simulated results of

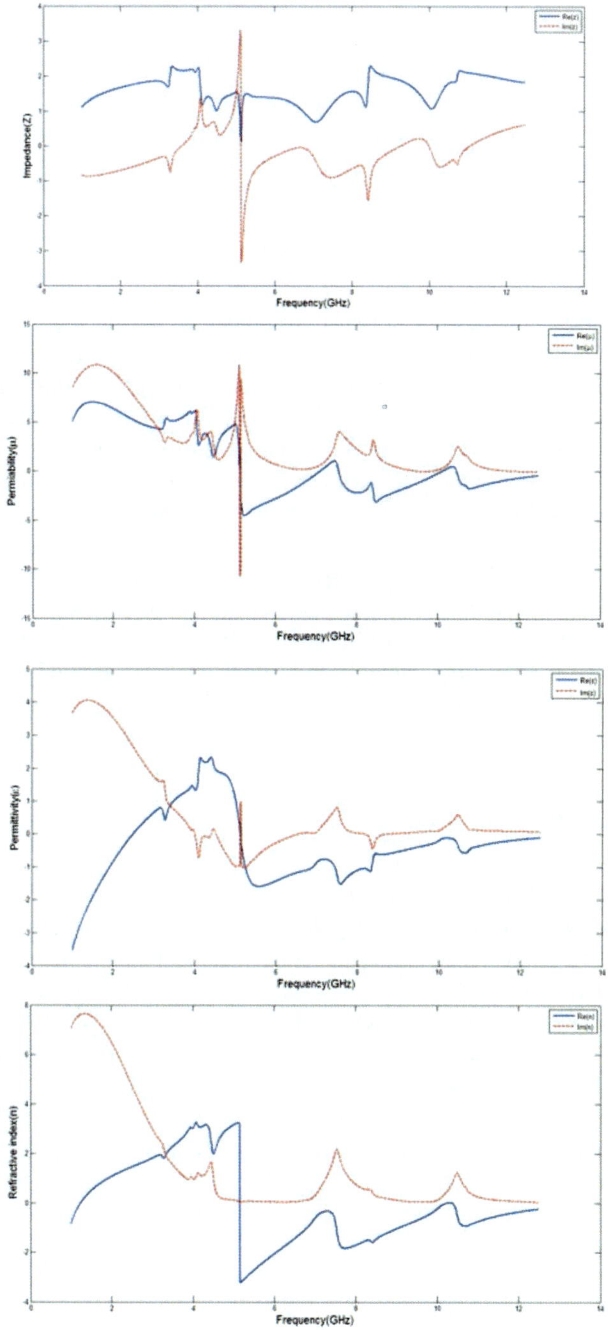

Fig. 7 Retrieved effective parameters of LHM

Table 4 Negative refractive index frequency bands of retrieved effective parameters

Parameter	Negative refractive index frequency bands (GHz)
Refractive index (*n*)	5.16–12.5
Permeability (μ)	5.2–7, 7.5–12.5
Permittivity (ε)	1–2.5, 5.2–12.5

proposed antenna directivity are measured as 2.5/1.8/ 2.2/ 2.9 and 2.6 dB at 4.7, 7.6, 8.8, 10 and 10.5 GHz.

Figure 10 shows the fabricated LHM multiband microstrip patch antenna using FR4 substrate with 0.8 mm thickness. Width and length of the radiating patch is $13 \times 13 \times 0.8$ mm^3 which is compact and less in weight. Figure 11 shows the simulation and measurement of proposed antenna. As can be seen, there is a significant agreement between the results of simulation and measurements, and the small differences are because of the effects of fabricated process and SMA connector. The return loss and VSWR of the proposed antenna is measured by Rohde & Schwarz ZNB 20 network analyzer. It is observed that the fabricated antenna has a return loss of −22.6 dB, −12.41 dB, −28.23 dB, −18.5 dB, −21.35 dB at 4.84 GHz, 7.6 GHz, 8.89 GHz, 10.06 GHz, and 10.83 GHz which are very close to the simulation result.

The functional parameters of the proposed antenna are compared with the reported antennas which are listed in Table 5. From the Table, it is examined that the prosed antenna performs better than the reported antenna.

4 Conclusions

The proposed LHM multiband frequency antenna was studied in this paper. The simulation of antenna is designed using Ansys HFSS v.15.0 simulation tool. Return loss, VSWR, gain, directivity, radiation pattern for both E—plane and H—plane are plotted. The efficiency of the proposed antenna is 95% at the frequency of 8.8 GHz which is considerable higher and also achieves UWB frequency range of 9.8–10.8 GHz. Fabricated antenna is measured using Rohde & Schwarz ZNB 20 network analyzer. Comparison between the results of simulation and measurements are made, and the small differences are because of the effects of fabricated process and SMA connector. The proposed antenna uses simple design and compact in size which operates at 4.74–10.56 GHz. The results obtained are found to be satisfying the requirements of the mid-range of 5G communication.

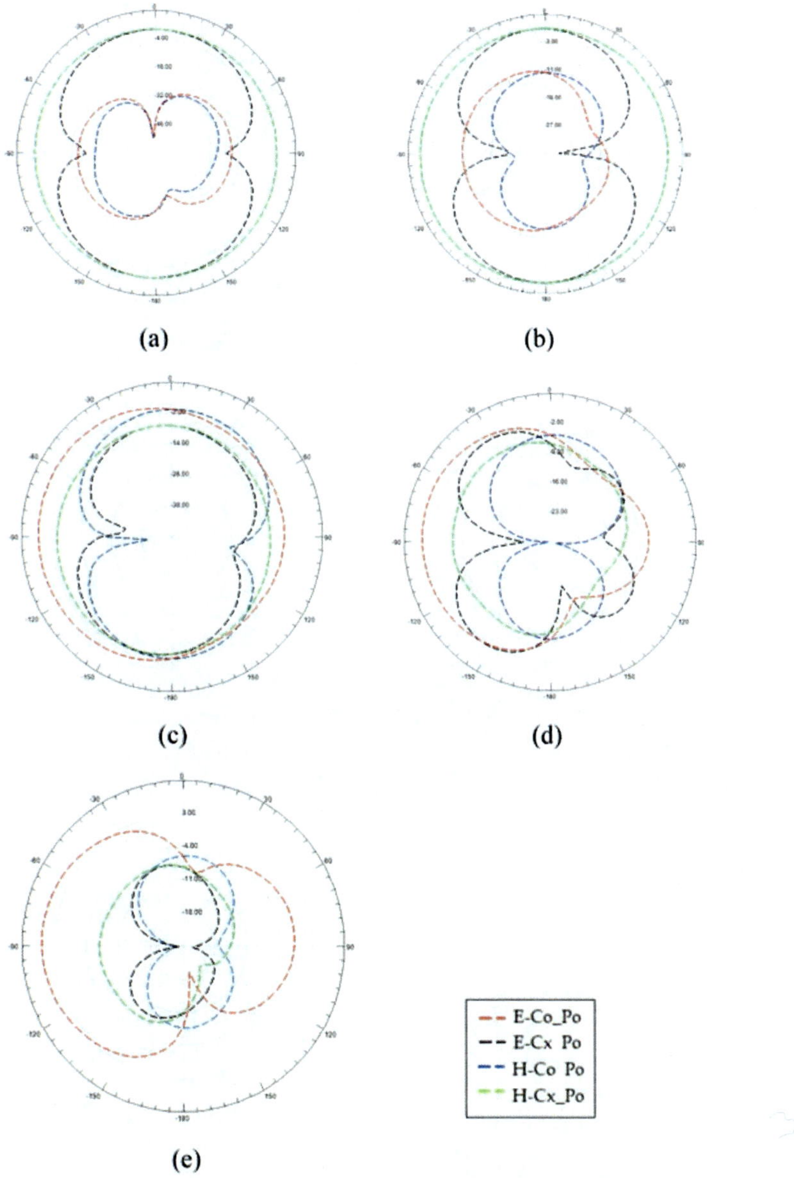

Fig. 8 The proposed antenna radiation characteristics at **a** 4.7 GHz, **b** 7.6 GHz, **c** 8.8 GHz, **d** 10 GHz and **e** 10.5 GHz

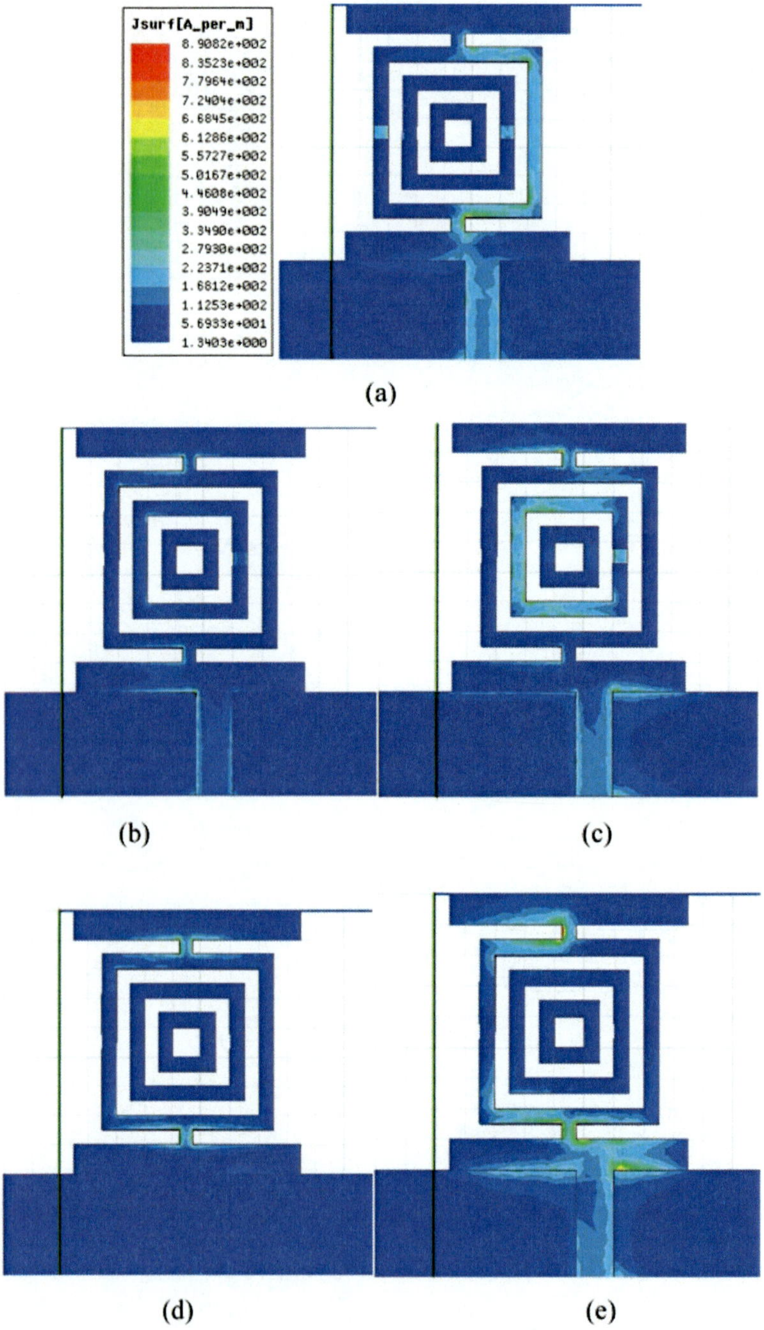

Fig. 9 Simulated current distributions of proposed antenna at **a** 4.7 GHz, **b** 7.6 GHz, **c** 8.8 GHz, **d** 10 GHz and **e** 10.5 GHz

(a) Fabricated antenna top view

(b) Fabricated antenna bottom view

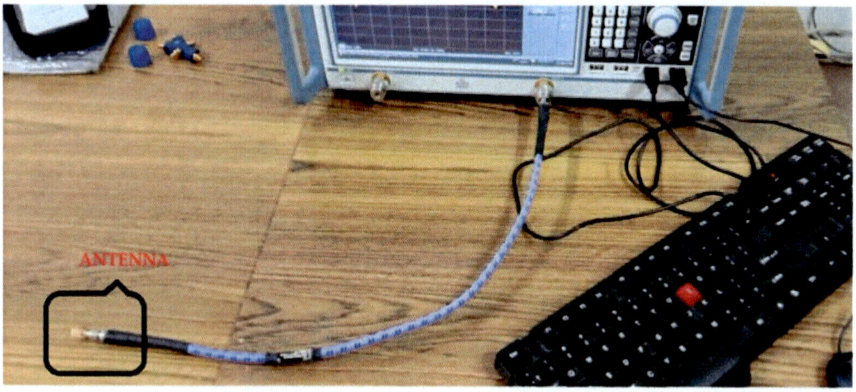

(c) Photography of measured setup of fabricated antenna

Fig. 10 Fabricated LHM multiband frequency antennas **a** Top view **b** Bottom view **c** Photograph of measured set of the antenna

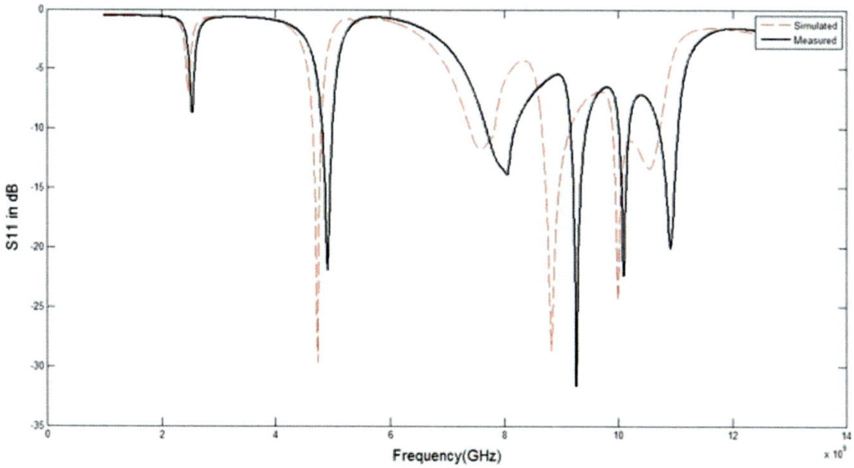

Fig. 11 Simulated and measured return loss of the proposed antenna

Table 5 Comparison between proposed and previously presented antenna

Ref. No	Dimensions (mm^2)	Thickness (mm)	Substrate	Frequency (GHz)	Gain	Return loss, S11	% size reductions
[12]	14 × 13.4	1.6	FR4	2.4	4 dB	−15.7 dB	1.65
[13]	50 × 50	2	FR4	3.1–10.6	8.9 dB	−20 dB	48
[14]	47 × 37	1.6	FR4	1.33 to 8.7	3.19 dB	−31.98	26
[15]	14 × 55	1.6	FR4	2.5	−1 dB	−32 dB	11
[16]	25 × 18	1.6	FR4	2.36 to 2.49 and 3.07 to 4.0	1.87 dB	−12.5 dB and −29 dB	5.8
[17]	45 × 25	1.57	Roger RT 5880	2.4 and 3.5	−2 dBi & 0.14 dBi	−26 dB and −22 dB	16
[18]	70 × 50	1.6	FR4	2.45 and 3.6	1.85 dBi	−19 dB and −29 dB	54
[19]	27.5 × 16.08	1.6	FR4	2.4, 5.2, 5.5, 5.8 and 7.4	0.37, 1.61, 1.88, 2.08, and 2.52 dB	−18, −17, −20, −13 and −10.05 dB	5.7
[20]	20 × 20	0.8	FR4	3–13.6	0–6 dB	–	1.8
Proposed antenna	13 × 13	0.8	FR4	4.7, 7.6, 8.8, 10, 10.5	1.2/ 1.3/ 2.1/ 1.2/ 1.4 dB	−22.6 dB, −12.41 dB, −28.23 dB, −18.5 dB, −21.35 dB	

References

1. Shelby RA, Smith DR, Schultz S (2001) Experimental verification of a negative index of refraction. Sci New Series 292(5514):77–79
2. Kafesaki M, ThKoschny, Penciu R S, Gundogdu T F et al (2005) Left-handed metamaterials: detailed numerical studies of the transmission properties. A pure Appl Optics 7(2)
3. Zhou X, Liu Y, Zhao X (2009) Low losses left-handed materials with optimized electric and magnetic resonance. Appl Phys 9:643–649
4. Smith DR, Padilla WJ, Vier DC et al (2000) Composite medium with simultaneously negative permeability and permittivity. Phys Rev Lett 84(4184–4187)
5. Wang J, Qu S, Zhang J et al (2009) A tunable left-handed metamaterial based on modified broadside-coupled split-ring resonators. Prog Electromagn Res Lett 6:35–45
6. Alici KB, Ozbay E (2008) A planar metamaterial: Polarization independent fishnet structure. Photon Nanostructures Fundam Appl 6(1):102–107
7. Burokur SN, Latrach M, Toutain S (2005) Theoretical investigation of a circular patch antenna in the presence of a left-handed medium. IEEE Antennas Wirel Propag Lett 4:183–186

8. Wang a J, Qu S, Xu Z et al (2009) Broadband planar left-handed metamaterials using split-ring resonator pairs. Photon Nanostructures Fundam Appl 7(2):108–113
9. EvrenEkmekci G-S (2009) Comparative investigation of resonance characteristics and electrical size of the Double-Sided SRR, Bc-SRR and conventional SRR type metamaterials for varying substrate parameters. Prog Electromag Res B 12:35–62
10. Marqu´es, Mesa R F, Martel J, et al (2003) Comparative analysis of edge- and broadside-coupled split ring resonators for metamaterial design-theory and experiments. IEEE Trans Antennas Propag 51(10):2572–2581
11. Marqu´es R, Mdina F, Rafii-El-Idrissi R (2002) Role of bianisotropy in negative permeability and left-handed metamaterials. Phys Rev B 65:14440(1)–14440(6)
12. Chinnathampy S, Aruna MN (2021) Antenna design: micro strip patch for spectrum utilization in cognitive radio networks. Wireless Pers Commun 119:959–979
13. Singh A, Mehra RM, Pandey VK (2020) Design and optimization of microstrip patch antenna for UWB applications using moth-flame optimization algorithm. Wirel Pers Commun 112:2485–2502
14. Jenath Sathikbasha M, Nagarajan V (2020) Design of multiband frequency reconfigurable antenna with defected ground structure for wireless applications. Wirel Pers Commun 113(2):867–892
15. Liua W-C, Wua C-M, Chub N-C (2021) A compact low-profile dual- band antenna for WLAN and WAVE applications. Int J Electron Commun 66(6):467–471
16. Mehedi Hasan MD, Faruque MRI, Islam MT (2019) Thin-layer dielectric and left-handed metamaterial stacked compact triband antenna for 2 GHz to 4 GHz wireless networks. J Electron Mater 48(6):3979–3990
17. Nandi S, Mohan A (2017) CRLH unit cell loaded Triband compact MIMO antenna for WLAN/ WiMAX, applications. IEEE Trans Antennas Propag 16:1816–1819
18. Abed AT, Singh MSJ (2016) Slot antenna single layer fed by step impedance strip line for Wi-Fi and Wi-Max applications. Electron Lett 52(14):1196–1198
19. Rajkumar R, Kiran KU (2017) A metamaterial inspired compact open split ring resonator antenna for multiband operation. Wirel Pers Commun 97(2):951–965
20. Kakhki MB, Rezaei P (2017) Reconfigurable micro strip slot antenna with DGS for UWB applications. Int J Microw Wirel Technol 9(7):1517–1522
21. Salai Thillai Thilagam J, Ganesh Babu TR (018) A Rectangular microstrip patch antenna at ISM band. In: Proceedings of the second international conference on computing methodologies and communication
22. Jahangiri M, Rajebi S (2020) Effects of split ring resonator (SRR) Metamaterial on the radiation pattern and variation of the heating focus points of the micro strip patch antenna. Bimonthly Peer-Rev J 5(1):307–316
23. Sumathi K, Lavadiya S, Yin PengZhi et al (2021) High gain multiband and frequency reconfigurable metamaterial superstrate micro strip patch antenna for C/X/Ku-band wireless network application. Wirel Netw 27:2131–2146
24. Yılmaz HO, Yaman F (2019) Metamaterial antenna designs for a 5.8GHz Doppler radar. IEEE Trans Instrum Meas 69(4):1775–1782
25. Rajak N, Chatoraj N, Kumar R (2019) A gain and bandwidth enhanced metamaterial based surface antenna for wireless communication. URSI Asia-Pacific Radio Science Conference
26. Abdelkebir S, Mayouf A, Mayouf F et al (2019) Study of metamaterial surface wave antenna based on split ring resonator. Microsyst Technol 25:797–810